Seeding Global Collaboration

EDITED BY

PATRICK BROWN

AND

JAMES DUFFY

Axial Publishing

Vancouver

Printed in Canada by
Grandview Printing Co. Ltd.
Vancouver, Canada
Email: info@grandviewprinting.com

Axial Publishing
www.axialpublishing.com

Canadian Cataloguing in Publication Data

Seeding Global Collaboration

ISBN   978-1-988457-00-0

1. Functional Collaboration   2. Methodology   3. Culture
I. Title

Text layout and cover:
Patrick Brown
James Duffy
Christina Ghorayeb

For

Esther Leilani Brown

and

Madison Elizabeth Duffy-Cyr

"The interpreter's differentiation of the protean notion of being must be not descriptive but explanatory. It will aim at ... the possibility of scientific collaboration, scientific control, and scientific advance towards commonly accepted results."*

"The modern sciences are not individual but community enterprises. They are not fixed achievements but ongoing developments. They are not isolated from one another but interdependent, usually in highly complex manners. The range of data to which they appeal and on which they rest is mastered, not by the individual but by the group, and not by the group of this or that moment but by the ongoing group that critically receives and independently tests each new contribution."†

"You are so young, so before all beginning, and I want to beg you, as much as I can, dear sir, to be patient toward all that is unsolved in your heart and to try to love the *questions themselves* like locked rooms and like books that are written in a very foreign tongue. Do not now seek the answers, which cannot be given you because you would not be able to live them. And the point is to live everything. Live the questions now. Perhaps you will gradually, without noticing it, live along some distant day into the answer."‡

---

* Bernard Lonergan, *Insight: A Study of Human Understanding*, ed. Frederick Crowe and Robert Doran, vol. 3, *Collected Works of Bernard Lonergan* (Toronto: University of Toronto Press, 1992), 609.
† Bernard Lonergan, "Variations in Fundamental Theology," *Philosophical and Theological Papers: 1965–1980*, ed. Robert Croken and Robert Doran, vol. 17, *Collected Works of Bernard Lonergan* (Toronto: University of Toronto Press, 2004), 245.
‡ Rainer Maria Rilke, *Letters To a Young Poet*, trans. M.D. Herter Norton (New York: W.W. Norton, 1934), 34–35.

Contents

"The most conspicuous effect of sustained investigation
and study is the division of labor."

Bernard Lonergan[1]

This volume ventures on a new path. It strives to illustrate and illuminate from various angles a profoundly innovative idea. That idea—a supremely practical but little-noticed idea—is functional collaboration. These essays collectively advance the claim that functional collaboration is the much-needed form for an efficient and productive division of contemporary academic and scientific labor, a method on the level of our times and proportionate to the complexity and breadth of modern modes of inquiry. The essays seek to foster, in other words, a "shift to the idea,"[2] where the idea is a method, and the method is functional collaboration. That method was the inspiration and driving force behind the Sixth International Lonergan Conference, "Functional Collaboration in the Academy: Advancing Bernard Lonergan's Central Achievement," held at the University of British Columbia in Vancouver, Canada, in July of 2014.

The papers presented at the conference, and published here, attempt to initiate a much-needed exploration and implementation of functional collaboration. The product of decades of effort, Lonergan's sketch of functional collaboration represents a richly suggestive and dynamic division of scientific and academic labor

---

[1] 69600DTEL60, 2. This sentence may be found in an early draft for chapter one of Bernard Lonergan's *Method in Theology* (New York: Herder and Herder, 1972). Like other archival documents cited in this volume, it may be found at the Lonergan Archives website: www.bernardlonergan.com.

[2] The German sociologist Georg Simmel coined the term, "*die Wendung zur Idee*," to refer to the shift towards system, the displacement towards the idea, the turn to reflection, gradually emergent and eventually verifiable in any body of sustained investigation and study. Chapter two of Simmel's *Lebensanschauung: Vier metaphysische Kapitel* (München/Leipzig: Duncker & Humblot, 1918) is titled "*Die Wendung zur Idee.*" The shift towards the idea and the reality of functional collaboration requires some form of implementation, however faltering or flawed that implementation may initially be. It is, after all, a method. Its latent potentialities have to be realized in a series of preliminary stages before it can mediate and manifest a fuller and more cumulative development, a shift towards system in Simmel's phrase and, in time, a shift towards genetic system. The latter shift is simply—or rather, complexly—Lonergan's articulation of genetic method applied to history as he himself intended. On the need for a genetic systematics, that is, a heuristic view on the geohistorical series of systems in any science or discipline, see generally the essays by Aaron Mundine and Meghan Allerton, as well as the Epilogue by Philip McShane.

into coordinated elements he calls functional specialties.[3] As a sophisticated and practical method for structured collaboration, his proposal has important and unexpected implications for the empirical sciences, the human sciences, and ultimately for the whole human struggle to solve the problems plaguing our societies, our economies, our institutions, our environment, and our world.

The key word here is "initiate." Each paper is a preliminary effort in a particular functional specialty in some specific context, together with an attempt to envision how the results of that functional effort might be handed on to the next specialty in the series, all cumulatively illustrating the functional specialties as constituent components of the broader and deeper method of functional collaboration limned so brilliantly and compactly by Lonergan. But these sketches fall far short of a fully-fledged intimation of that method; they are manifestly inadequate versions of the more complex and profound idea articulated by Lonergan, beta versions in search of better versions. Still, they are not without value. The contributors to this volume are reading the method from the point of view of implementation, as any method demands. A method is made to be implemented, not sequestered under a bell jar somewhere. According to Collingwood, the visual artist does not paint what she sees; she paints in order to see better.[4] Similarly, the contributors are not sketching what they read, so to speak; they are sketching in order to read better. On the individual essays, more will be said below. But first we provide an initial orienting context.

*   *   *   *   *

Any modern science is conspicuously and relentlessly collaborative. Scientific investigation and scientific results are the continuing achievements of a very large group, and a highly specialized group at that. Some sense of the depth and scope of this collaboration may be attained simply by glancing at the scientific paper announcing the verification of the Higgs Boson in 2012. The effort to verify the hypothetical particle involved extensive work at the Large Hadron Collider,[5] and the verification eventually helped Peter Higgs earn the Nobel Prize in Physics the following year. No less than sixteen single-spaced, small-print pages of the published paper are entirely devoted merely to naming the co-authors and collaborators

---

[3] The eight functional specialties are research, interpretation, history, dialectic, foundations, systematics, policies (or doctrines), and communications. The present volume presupposes readers who have already made serious or sustained efforts to understand the specialties. Any reader who is intrigued by these essays but not yet familiar with Lonergan's delineation of the specialties should consult *Method in Theology*, especially chapter five.

[4] R.G. Collingwood, *Principles of Art* (Oxford: Oxford University Press, 1938), 303–04.

[5] For background information, see https://home.cern/topics/large-hadron-collider.

involved in the discovery, over 3,000 contributors from over 170 institutions around the world.[6]

The reach of any contemporary science is far beyond the grasp of any individual scientist, and the same could be said of any specialty or subspecialty within an established science. Empirical science operates on the basis of a vast collaboration of many individuals over time, and the collaboration is continuous, cumulative, massive, organized, and increasingly global. In other words, a science is the product of a large group continuously operating within a *de facto* structure for collaboration called scientific method, a collaboration that yields cumulative and progressively revised and refined results.

Put simply, empirical science is a radically collective enterprise.[7] A scientific group divides its total labor among specialties, distributes the various efforts to different specialists, conducts its complex investigations, and eventually pools and communicates the results. The cycle is then repeated, over and over again. Nor are the empirical sciences unique in being the product of a community enterprise, a collaborative group effort. "Any science, any academic enterprise, is the work of a group, of a scientific or academic community. For the work to prosper, the conditions of its possibility must be fulfilled."[8]

What are those conditions? The received wisdom might find very strange the assertion that identifying and fostering the emergence of those conditions in a precise and profound way occupied Bernard Lonergan for decades. But of that there can be no doubt. "It is with the conditions, preliminary to an effective collaboration, that the present work is concerned."[9] Lonergan's efforts to articulate and implement the conditions for effective and efficient collaboration were not limited to his long labors to articulate a *Method in Theology*. The "present work" in the sentence quoted above refers to the whole of the book *Insight*, and "the conditions, preliminary to an effective collaboration" may be thought of as the conditions required for the work of future scientific and academic communities to prosper, conditions he later spelled out in the basis and framework for collaboration he called functional specialization.

---

[6] "Observation of a new boson at a mass of 125 GeV with the CMS experiment at the LHC," *Physics Letters B*, 716 (2012), 30–61. For the list of contributors and their institutions, see *ibid.*, 45–61.

[7] Bernard Lonergan, *Insight: A Study of Human Understanding*, ed. Frederick Crowe and Robert Doran, vol. 3, *Collected Works of Bernard Lonergan* (Toronto: University of Toronto Press, 1992), 727 (hereafter *Insight, CWL* 3).

[8] Bernard Lonergan, "Variations in Fundamental Theology," *Philosophical and Theological Papers: 1965–1980*, ed. Robert Croken and Robert Doran, vol. 17, *Collected Works of Bernard Lonergan* (Toronto: University of Toronto Press, 2004), 253.

[9] Bernard Lonergan, "The Original Preface of *Insight*," *METHOD: Journal of Lonergan Studies* 3:1 (1985), 3–7, 6.

What, then, are the conditions required for scientific or academic work to prosper? First, the workers in every field must be keyed to the *need* for efficient collaboration. In earlier and more primitive times, perhaps, individual scientists and scholars had the luxury of "singly following the bent of their genius, their aptitudes, and their acquired skills."[10] But as sciences develop and differentiate, as fields multiply and divide, as complexity increases and knowledge is distributed in ever more intricate patterns throughout the relevant groups, merely individual work is less and less tenable and less and less valuable. Instead, those laboring within any science are forced "to collaborate in the light of common but abstruse principles and to have their individual results checked by general requirements that envisage simultaneously the totality of results."[11] The same may eventually be expected of academic work, once the academy finally slips the traces of its outmoded and outdated individualist mentality. Unfortunately, that almost inveterate mentality survives to generate distrust of Lonergan's elegant solution to the problem of efficient collaboration. It's hard to embrace a solution to a problem one does not even know exists.

Second, the workers in every field of scientific and academic endeavor must be keyed to some *method* for efficient collaboration. The "common but abstruse principles" in light of which they are to collaborate must be in some sense shared, held together and held in common as a guiding goal. How, though? Initially, and cyclically, the principles may be held and shared "by comprehending in a unified whole all the conclusions intelligibly contained in those very principles," a comprehending that "cannot take place without a construct of some sort."[12] The relevant construct is a complex image or diagram, or rather a strategic series of them. "In larger and more complex questions, it is impossible to have a suitable phantasm unless the imagination is aided by some sort of diagram. Thus, if we want to have a comprehensive grasp of everything in a unified whole, we shall have to construct a diagram in which are symbolically represented all the various elements of the question along with all the connections between them."[13]

One may think here of Philip McShane's heuristic and strategic meta-words or meta-grams.[14] Though they may strike an overly conventional Lonergan scholar as unacceptably unconventional, the meta-words are themselves the relatively straight-

---

[10] *Insight, CWL* 3, 604.

[11] *Ibid.*, 604.

[12] Bernard Lonergan, *The Ontological and Psychological Constitution of Christ*, ed. Michael Shields, Frederick Crowe, and Robert Doran, trans. Michael Shields, vol. 7, *Collected Works of Bernard Lonergan* (Toronto: University of Toronto Press, 2002), 151.

[13] *Ibid.*, 151.

[14] Philip McShane, *"Prehumous* 2: Metagrams and Metaphysics," http://www.philipmcshane.org/prehumous; see also McShane, *A Brief History of Tongue: From Big Bang to Coloured Wholes* (Halifax: Axial Press, 1998), 108–110, 119, 122, 124.

forward product of Lonergan's own explicit stances. They reflect, for example, Lonergan's stance regarding the significance of adequately heuristic symbolism,[15] Lonergan's view that explicit metaphysics consists "in a symbolic indication of the total range of possible experience,"[16] and Lonergan's insistence that for adequate explanatory investigation of the multi-layered human organism, "there have to be invented appropriate symbolic images of the relevant chemical and physical processes,"[17] the relevant botanical and zoological processes, and the relevant intellectual and religious processes. With the help of such schematic and heuristic symbolic images, scientists investigating human reality gradually can identify the regularities, laws, realized schemes, and capacities-for-performance of each layer in its interactions with other layers in the dynamic system that is a human organism— to say nothing of the cosmic, evolutionary, historical, and linguistic processes in which the layers emerge and develop.

Third, one of the conditions for scientific or academic labor to prosper is a startling shift in viewpoint and in sensibility that may helpfully be called "the heuristic turn." That shift, or turn, is at one level a development and an elaboration of the need for complex heuristic images emphasized by Lonergan and taken for granted by virtually every community within the empirical sciences. That turn is, if you prefer, a basic implication and revelation of the heuristic devices and desires[18] threading through any science. Think of the Periodic Table in chemistry, or in the human sciences, think of Lonergan's complex diagram of "the skeletal structure of all human process."[19] A "principle," according to Aristotle, is what is first in an ordered set, and complex heuristic images in that sense are principles. However abstruse they may initially appear, the needed complex images must be, in some sense, first held in common by the group, if the images are to heuristically and therefore successfully represent and re-present all the elements of collaboration and the connections between them, as well as guide collaborative investigations.

We mentioned the need for a radical shift in "sensibility" as well as viewpoint. It is worth pausing on this point. Perhaps it is caused by the ghostly phantom of a

---

[15] See generally, Lonergan, "The Significance of Symbolism," *Insight, CWL* 3, 42–43, and see notes 66 and 67 below.

[16] *Insight, CWL* 3, 421.

[17] *Insight, CWL* 3, 489. Robert Henman's essay in this volume elaborates on this theme.

[18] We allude here to the famous phrase from the *Book of Common Prayer*. "The original sense of device was 'desire, intention.' ... The source of device is a French form based on Latin *dividere* 'to divide.' Its sense developed from 'desire, intention' to 'a plan, scheme, trick' and then the usual modern meaning of 'a thing made or adapted for a particular purpose.'" Julia Cresswell, *Oxford Dictionary of Word Origins* (Oxford: Oxford University Press, 2004), 125.

[19] Lonergan, "Finality, Love, Marriage," *Collection*, ed. Frederick Crowe and Robert Doran, vol. 4, *Collected Works of Bernard Lonergan* (Toronto: University of Toronto Press, 1988), 41– 42, and see also Michael George's essay below.

residual Platonism, or perhaps it is rooted in some other cause, but why would we not expect a painful challenge to present sensibilities presented by Lonergan's proffer of a radically innovative method? "The greater the novelty, the less prepared the audience."[20] Is it not foolish to presume that we are somehow already prepared for that great novelty? Is it not foolhardy to assume that our established and inertial routines would somehow not get the better of us? Given the novelty of functional collaboration, should we not reasonably expect instead that we are fundamentally unprepared for its fuller meaning, unprepared not only on the level of mind, but also on the level of psyche, of sensibility, unprepared on the level of antecedent willingness? For example, if your settled expectations as you begin to struggle with the complexities of functional collaboration are roughly those of a *littérateur* confronting a work in advanced physics, insisting that physics somehow take the trouble to translate itself into evocative literature on your current level of capacity-to-apprehend, then you will be among the audiences who are radically and psychically unprepared for the required shifts.

The new method calls for "fresh adaptations"[21] of sensibility, and those will not come readily or easily to the unprepared and the unwilling. Indeed, it would seem fatally easy to underestimate the magnitude and complexity of what is called for by either of those two words. The developments otherwise due, the needed "fresh adaptations" that should occur, may not in fact occur, may in fact be passively or even vehemently resisted. Or if they do occur they may be mangled, or they may occur only in scattered members of the community, who then will be promptly shunned by the pre-adapted majority in the community. Think in this context of the fates of Socrates and Jesus; think of the various beatings meted out to Paul of Tarsus as recounted in *Acts*; think of "the birth trauma of modern science"[22] out of a dominant and decadent Aristotelianism; think of Lonergan's penetrating analysis of "the tension of community."[23] Think even of the checkered reception-history of Lonergan himself.

Fourth, the only real alternative to something like the heuristic turn is a kind of decadent nominalism, one that shirks the pain and labor of explanation and settles instead for comfortable and casual description. But that is ultimately a form of the

---

[20] *Insight*, *CWL* 3, 612.

[21] *Insight*, *CWL* 3, 248 ("Human sensitivity is not human intelligence, and if sensitivity can be adapted to implement easily and readily one set of intelligent dictates, it has to undergo a fresh adaptation before it will cease resisting a second set of more intelligent dictates.") On antecedent willingness, see *ibid.*, 633, 646, 711.

[22] 18490DTE060, 12 (lecture notes titled "Horizons and Categories" for the 1968 institute on method in theology at Boston College); Bernard Lonergan, *Early Works on Theological Method 1*, ed. Robert Doran and Robert Croken, vol. 22, *Collected Works of Bernard Lonergan* (Toronto: University of Toronto Press, 2010), 488.

[23] *Insight*, *CWL* 3, 239–44; see especially *ibid.*, 559.

flight from understanding, perhaps even what Lonergan in another context called the "treacherous, insidious ... refusal of the explanatory viewpoint."[24] By the same token, it is not easy to take up the challenge. Far from it. The challenge of the explanatory viewpoint—the challenge, for example, of concrete and detailed historical and dialectical expansions of the virtual and heuristic viewpoint on viewpoints Lonergan develops in *Insight* and, so to speak, distributes in *Method*—is profoundly difficult, a mighty and weighty challenge indeed.

Fifth, the heuristic turn brings out the importance of a standard model. Physicists can collaborate efficiently because they have developed, and can rely on, a standard model, a heuristic structure coordinating diverse and far-flung investigations. The standard model is not fixed and immutable; it shifts and changes to some extent with every major breakthrough. That is to say, the model itself develops, even while it guides developments and advances in the contemporary state of the science physics. The model itself makes possible sustained and efficient collaboration; the collaboration yields cumulative and progressive results and revisions; and those results and revisions in turn lead to further modifications and refinements of the standard model. Nothing like that complex, elaborately choreographed, and immensely productive ballet of collaboration and coordination exists in the human sciences because the human sciences as of yet have nothing comparable to the standard model in physics.[25]

Sixth, the mighty and difficult challenge of functional collaboration extends to all sciences—think in particular of contemporary physicists climbing beyond the doing of research and interpretation within a standard model. Their story is to be transformed by the remarkable claim that "theoretic understanding, then, seeks to solve problems, to erect syntheses, to embrace the universe in a single view,"[26] by a luminous handling of the significance of the distinction between primary relations and secondary determinations, [27] and by an effective dealing with myths of reductionism and popularizability regarding the full progressive geometry of physical

---

[24] 41700DTE050, 16 (header 81) (ms. for chapter 17 with changes)(out-take from *Insight*, *CWL* 3, 563). The typed paragraph from which the quotation is taken is crossed-out in the manuscript.

[25] Constructing the basis for an adequate standard model for the human sciences was, arguably, Lonergan's life-work. It would certainly seem difficult to read *Method*, 286–87 in any other context or, for that matter, his contention that *Insight* was centrally concerned with creating and fulfilling "the conditions preliminary to an effective collaboration." "The Original Preface of *Insight*," METHOD: *Journal of Lonergan Studies*, 3 (1985), 6. One must not, of course, commit the blunder of imagining human sciences as presently configured merely retrofitted with a standard model, "new and improved," so to speak. They will be changed, changed utterly.

[26] *Insight*, *CWL* 3, 442.

[27] *Insight*, *CWL* 3, 515–17.

realities in their secondary relativities.[28] An effective sub-task of the functional collaboration for scientists will be to live in a concrete dialectical distinguishing of predominantly methodological issues from those issues flowing mainly from under-developments in understanding the physics of history.[29]

Seventh, the vague discomfort you may have felt while reading the last paragraph suggests a final condition for scientific or academic labor to prosper. That condition is a personal and group willingness and readiness to risk a new view, to venture on a new path, to upend mere conventionality where mere conventionality serves only to stifle that lively and probing spirit of inquiry on which all intellectual flourishing so vitally depends. But stifling conventionality does not always come with a label attached. It may be so familiar as to be virtually unnoticed, so widely shared as to seem beyond question. It can take the form of a comfortable complacency with the usual norms or the usual views. Already accepted procedures may take on a life of their own simply because they are already accepted. Familiarity may breed neglect, or worse, comfortable closure.

"Some kinds of work, no doubt, occur within a horizon open to expansion; and expansion can and does occur. But it is rare that openness is built into living and working, rare that its fruit is esteemed by many, rare that much significance is attached to the meanings and values that would change customary ways of life."[30] Customary ways of life, or non-life, are rife in the academic world, largely taken for granted and so unnoticed. Through its cyclic and cumulative operating, functional collaboration will gradually structure a currently-rare kind of openness and expansiveness into living and working in scientific and academic fields, and from there into wider fields of culture, broader and deeper realms of living and working.

However brief and incomplete, these seven themes may suffice to provide readers of this volume with a preliminary idea of the value and the challenge, the prospects and the perils, of the revolution in method that is functional collaboration. The essays themselves fill out and expand this preliminary sketch.

* * * * *

---

[28] See Philip McShane, "Elevating *Insight*: Space-Time as Paradigm Problem," *METHOD: Journal of Lonergan Studies*, 19 (2001), 203–229.

[29] "Such dialectic work, in a full specialist context, would take its accurate place in the spiral of specializations that begins and ends with Communications, grounding a lift in the probabilities of, for example, good teaching, good popularization, and good technology." McShane, *ibid.*, 216.

[30] 29810DTE070, 5 (header 11)(fragment from a draft of Lonergan's article, "Horizons and Transpositions").

The contributions to the 2014 conference were drafted and circulated prior to the conference. The authors agreed on a common format for the essays, one in which each author (1) identifies a context, his or her current understanding of the task and why he or she thinks it is worth the effort; (2) attempts to specify a "content," something that, if cycled forward, might transform the collaborative effort and indeed transform a concrete situation such as a school or school board, a local economy or polity, a church community or other local institution; (3) attempts to hand on efficiently the relevant content to an audience; (4) reflects critically on what was learned in the prior two attempts to think and write functionally. This four-part format allowed each author to make two identifiable efforts at functional writing bracketed by reflections about his or her efforts.

Most essays in this collection are loosely identifiable as pertaining to a particular disciplinary zone. In addition, each author focuses on a specific collaborative task, with the exception of Meghan Allerton, whose essay explores intervention and effective address in the area of ecology in the light of the functional specialties in their intrinsic relations to one another.

In his essay, "Functional Research in Neuroscience," Robert Henman zeroes in on the key task of functional research and on key texts from chapters 15 and 16 of *Insight* that are to be components of an effective ethos of progress. The key task is to find something that deserves recycling through the structure of functional collaboration. The key texts are (a) "there have to be invented appropriate symbolic images" [31] if one is studying organic, psychic, and intellectual developments empirically and integrally, as triply compounded, and (b) "there results the problem of formulating the heuristic structure of the investigation of this triply compounded development."[32] Generalized empirical method poses a novel challenge to philosophers, neuroscientists, and all those striving to move beyond description and reductionism towards an adequate explanatory account of the relationships between the brain and the mind alike. When the philosopher of science or neuroscientist studies the nervous system without an adequate heuristic structure for his or her investigation, what results is "pseudometaphysical mythmaking."[33]

Terrance Quinn attempts the exercise of functionally interpreting chapter 5 of *Insight*, "Space and Time." He focuses on two theorems regarding (a) the abstract formulation of the intelligibility immanent in Space and in Time, and (b) the concrete intelligibility of Space that grounds the possibility of those simultaneous multiplicities named situations. Quinn notes that understanding these two theorems is no mean task, for it asks the reader to move beyond the series of exercises in self-attention in the first four chapters of *Insight*. Understanding the first theorem leads to a provisional acceptance on the part of a community of physicists who have done

---

[31] *Insight, CWL* 3, 489.
[32] *Insight, CWL* 3, 494–95.
[33] *Insight, CWL* 3, 528.

the prior exercises, so these pointers from *Insight* are doctrinal in the sense that they point to being luminous about, and within, a particular genus of development within physics. Quinn's interpretation of the second theorem begins with a question regarding the situations on the night of *An Oriental Monsoon*, a performance by a dance troupe from Hangzhou Normal University in China. He suggests that the physics of performers, performance, and audience that is, conceivably, the concern of a physics community, is also an invitation to be luminous about, and within, the dynamic totality that is emergent probability.

Patrick Brown's essay is a foray into some of the wider issues involved in interpreting Lonergan, including the problem of differentiated expression. In that sense, his essay may be viewed as providing central and helpful leads to interpreting the other essays in this volume. But the essay also provides a limited interpretation of a brief but complex passage in a letter Lonergan wrote to Frederick Crowe in 1954. That letter contains a remarkable paragraph in which Lonergan discusses how his view of "the Method of Theology" was "coming into perspective" by using a complex mathematical analogy. For Brown, an adequate interpretation of this paragraph would move beyond contemporary conventional expectations, expressions, and scholarly procedures—and would in fact ultimately require the context of developed functional collaboration. In his effort to functionally interpret the meaning Lonergan had in mind for the technical phrase, "evaluating $[1 + 1/n]^{nx}$ as $\underline{n}$ approaches infinity," Brown suggests a proximate context for what was "coming into perspective" and suggests why Lonergan may have selected that equation to express his growing perspective. But his interpretation focuses mainly on what the text does *not* mean. A more positive and technical interpretation of the equation appears in an "Interlude" by Phil McShane. In addition to interpreting the equation, McShane takes up the baton of interpretation as a functional historian would and locates Lonergan's 'coming into perspective' in two dense paragraphs, which he likens to a two-paragraph summary of Andrew Wiles' handling of Fermat's Last Theorem. Brown then attempts to read those two paragraphs as incarnating both "the heuristic turn" and a meta-scientific level of expression.

In "Functional History and Functional Historians," Aaron Mundine homes in on the functional historian's task of revising the genetic sequence of effective meaning, or meaning that actually impacts local situations, once functional collaboration has developed. He draws on a lengthy passage from *"De Intellectu et Methodo"*[34] in which Lonergan claims that the historian of any discipline has to have a thorough knowledge and a systematic understanding of the whole subject in order to write an adequate history of the subject. That passage pivots on an analogy

---

[34] Bernard Lonergan, "Understanding and Method" (*De Intellectu et Methodo*), trans. Michael Shields, in *Early Works on Theological Method 2*, ed. Robert Doran and H. Daniel Monsour, vol. 23, *Collected Works of Bernard Lonergan* (Toronto: University of Toronto Press, 2013), 175–77.

between the developing set of insights in the historian of a discipline, who is necessarily an expert in the discipline whose history she writes, and the parallel developed and heuristic view that would be required to understand the broader historical process. Mundine examines how Aristotle's position on the human mind understanding the intelligibility in an imaginative or sensible presentation might be received and passed on. Hypothetical functional historians receive fine interpretations of those who have written on the topic of "insight into the phantasm" from hypothetical interpreters who are at home in generalized, empirical neuroscience and pass on to the hypothetical dialecticians a revised genetic sequence of the study of human understanding. The hope is that this will give a lift not only to the specific tasks of functional dialectic but eventually also to varieties of common sense and more broadly to the schools and classrooms and educational structures of the future.

James Duffy's essay is an exercise in identifying the foundational reality of a special personal relation. He focuses on a key paragraph in *The Triune God: Systematics*[35] and on *Cherishing*,[36] the name McShane gives to the *esse secundarium* of the incarnation, a cosmic base and "an alliance and a love that, so to speak, bring God too close to man [and woman]."[37] The problem of external and internal relations is taken up by Lonergan in chapter 16 of *Insight*, and Duffy claims that if you want to get to grips with the meaning of the secondary act of existence of the incarnation, a decent place to start is section 2 of this chapter and Appendix 3 of *The Triune God: Systematics*. In his essay he intimates the novelty of functional collaboration by questioning whether and how poets and mystics contribute to foundationalizing cumulative and progressive results. In an appendix to his essay, in the spirit of self-assembly, Duffy reflects upon formative moments in his journey.

In his essay, "The Fifth Functional Specialty and Foundations for Corporate Law and Governance Policies," Bruce Anderson searches for foundations for policies concerning corporate governance. His guiding question is, "What is the ground for doctrines about how to intelligently govern and manage corporations?" The context for his search for a foundational perspective can be identified descriptively with the differentiations that are listed in *Method in Theology* on pages 286–87. Anderson holds that a basic knowledge of economics is necessary to ground good corporate governance. He provides a highly compact version of Lonergan's explanatory perspective on how the parts of an economy fit together, as well as another compact formulation of norms for good corporate governance that

---

[35] Bernard Lonergan, *The Triune God: Systematics*, trans. Michael Shields, ed. Robert Doran and Daniel Monsour, vol. 12, *Collected Works of Bernard Lonergan* (Toronto: University of Toronto Press, 2007), 471–73.
[36] See Philip McShane, *Posthumous* 12, "Clasping, Cherishing, Calling, Craving, Christing." This essay is available at: http://www.philipmcshane.org/posthumous.
[37] *Insight*, CWL 3, 747.

would follow the norms of a properly functioning economy. An example of a fundamental doctrine would be the replacement of the shareholder primacy model of corporate law and governance by a position on money and profit that emerges from understanding how economies actually function. Anderson notes that a foundational communication of this doctrine to commonsense scholars with an interest in corporate governance would face the difficult task of communicating the basics of sane economics to those who are neither theoretically inclined nor focused on immediately practical solutions.

Sean McNelis uses housing as a case study to attempt the functional specialty doctrines, or what he prefers to call policies. He identifies various dimensions of housing, including environmental, technological, economic, political, cultural, and religious dimensions. But he notes that one pervasive context for the mess in housing policies may be found in dominant dynamics—themselves the results of past policies—of wealth accumulation or profit maximization. Drawing upon an analogy with the team of Dr. House, who bring an increasing and so changing understanding of various systems and sub-systems to any diagnostic situation, McNelis claims that the disease of housing lies within one sub-system among a number of sub-sub-systems. A collaborative treatment would include a foundational decision to appropriate and implement explanatory definition, to integrate diverse disciplines and diverse methods used in housing research, and to implement a new heuristic of the economy. Foundational fantasy would speculate on what might emerge from an appreciation of human capacity to understand, to create, and to dwell, to be fully in the world.

In his essay, "Identifying the Eighth Functional Specialty," William Zanardi focuses on the problem of functionally communicating the non-reception of functional specialization. His concern is to effectively persuade others, especially new students, to exploit the possibilities of functional specialization. He identifies the fear of being displaced, or 'homeless,' as an impediment to seeding functional collaboration. Focusing on the phrase, "identification is performance,"[38] and taking up the analogy of producing cars—while many hands go into the making of cars, it is marketing and sales that make the earlier labor pragmatically purposeful—Zanardi endorses a series of initial experiments in functional communications that will be both mindful of linguistic feedback and hopeful that learning will follow doing. There can be no beginning implementation without the messiness of trial and error and initially mixed results. Zanardi believes that functional specialization, even in its not-yet-mature stages, will produce performances transforming the conventional academic fare.

The larger context of Philip McShane's essay, "Foundations of Communications," includes his many years spent struggling with the problem of

---

[38] *Insight, CWL* 3, 582.

communication, a problem he first encountered when he came across the problem of Cosmopolis in the late 1950s. In the 1980s he began to seriously struggle with chapter 14 of *Method in Theology*, and the outcome of that effort was his seminal article, "Systematics, Communications, Actual Contexts."[39] Other writings on the same problem include "Communications and Ever-ready Founders"[40] and "Structuring Systems in Towns, Gowns, and Clowns."[41] In his contribution to the present collection of essays, McShane weaves reflections about the kataphatic fantasy of foundations persons with ruminations on the poetry of Hafiz. The key insight regarding the foundations of communications is "seeing now, smelling now, each city block or rural farm as under the umbrella of an eight-layered towering collaboration of situations."[42] McShane notes that poetic positioning is part of each person's struggle for his or her own integral luminous consciousness.

In "Communicating Macroeconomic Dynamics Functionally," Michael Shute focuses on the difficulty of communicating macroeconomic dynamics both functionally and effectively given the current situation in which neither general method nor economics are operable sciences, or achievements of communal meaning. Two difficult challenges emerge. The first is to understand how a smooth transition from the surplus expansion to the basic expansion might happen, and the second is to figure out how to communicate such an understanding to heterodox economists. Shute provides various examples of attempts to "hand-on" through correspondences. One is a letter he wrote to an institutional economist who had written an article critically assessing *Lonergan's Discovery of the Science of Economics*.[43] The second is an email exchange with a documentary film-maker who has proposed doing a documentary on Lonergan. Shute adds an extended blogpost as an example of outreach, and concludes that functional communicating in economics is a tremendous creative challenge to envision effective ways to intervene in situations that "are the cumulative product of previous actions ... guided by the light and darkness of dialectic."[44]

Michael George focuses his essay on Lonergan's article, "Finality, Love, Marriage," first published in *Theological Studies* in 1943. As Fred Crowe points out in

---

[39] Philip McShane, "Systematics, Communications, Actual Contexts," in *Lonergan Workshop*, ed. Fred Lawrence, vol. 6 (Atlanta, GA: Scholars Press, 1986), 143–74.
[40] *Cantower* 14 (May 1, 2003), available at: http://www.philipmcshane.org/cantowers.
[41] This is the title of chapter 14 of *Futurology Express* (Vancouver: Axial Publishing, 2013), 92–98.
[42] See below, page 169.
[43] Michael Shute, *Lonergan's Discovery of the Science of Economics* (Toronto: University of Toronto Press, 2010).
[44] *Method in Theology*, 358.

his notes to the *Collected Works* version of the article,[45] Lonergan's analysis of love was neither expanded nor developed in his later work. George suggests there was more going on in this article than a mere justification and reiteration of the traditional teaching on marriage, and it is imperative to recover and recycle issues of sexuality in terms of a normative dimension of growth and development. The particular text that he strives to cycle forward is compact but deeply significant: "The ignorance and frailty of fallen man tend to center an infinite craving on a finite object or release: that may be wealth, or fame, or power, but most commonly it is sex."[46] In his essay George concentrates on foundations and its *per se* function of fantasizing. Foundational beginnings of normative reflection on sexuality would consider the differentiation of terms and relations implicit in the general theological categories on pages 286–287 of *Method in Theology*.

Meghan Allerton's essay explores effective address and intervention in the light of functional specialties in their intrinsic relations to one another. Drawing upon the work of Stewart Brand, she explores how research using high technology might be linked to policy-making in the area of ecology. She sees, perhaps surprisingly, that ecology faces the same challenge that theology faces, i.e., to differentiate and order a sequence of tasks that have a pragmatic bent. Allerton notes that the ecological movement has, from the beginning, taken a pragmatic turn. She identifies an emerging need to transpose descriptive histories of both decent interpretations and misinterpretations of deforestation and drilling for oil into the fuller context of scientific interpretation intimated in the third section of chapter 17 of *Insight*. She writes that the break from classical culture's ideal of "one solution fits all," perhaps more evident in ecology than in theology, requires seeding a geohistorical systematics.

* * * * *

What each reader will glean from reading the essays in this collection, will depend upon his or her horizon of academic formation, his or her beliefs, interests, needs, and desires. Some readers will be students striving to get a degree and move on to the next stage, while others will be teachers struggling to give bread and not stones to high school students, undergraduates, or graduate students. In either case, given

---

[45] "Finality, Love, Marriage," *Collection*, ed. Frederick Crowe and Robert Doran, vol. 4, *Collected Works of Bernard Lonergan* (Toronto: University of Toronto Press, 1988), 17–52.
[46] "Finality, Love, Marriage," *CWL* 4, 49.

both the non-relaxed pace [47] and the "the barbarism of specialization" [48] —a barbarism that determines who is suitably qualified to teach high school students, undergraduates, and graduate students, and what is suitable for publication— effectively intervening in academic culture is and will continue to be messy. Seriously asking how the foundations and doctrines of housing dynamics might be changed given the stronghold of past policies can be a mind-boggling experience.[49] We could ask the same question regarding possibilities of transforming the dynamics of lower and higher education. Similarly, how might research using high technology be linked to ecofriendly policy-making?[50] It is probably best to divide up these and other needs to link.[51] But how, exactly?[52]

A glance at the table of contents reveals a spread of topics including neuroscience, space and time, corporate law, housing, macroeconomics, Lonergan's 1943 essay on the ends of marriage, his 1954 perspective on the method of theology, and methodological reflections on current ecological crises. So varied a set of topics might seem random or unconnected. But they are not. Four factors combined to create a shared meaning and strategy in the community of contributors attempting to advance functional collaboration in this volume and to seed its future practice.

First, the authors share the conviction that functional collaboration is a *novum organon*,[53] a new and highly significant contribution to the collective possibilities of

---

[47] See Aaron Mundine's essay at pages 81–82.

[48] The title of chapter 12, José Ortega y Gasset, *The Revolt of the Masses*, trans. Anthony Kerrigan, ed. Kenneth Moore, with a Foreword by Saul Bellow (South Bend, IN: University of Notre Dame Press, 1985).

[49] See Sean McNelis's essay, the subsection entitled, "Prior to Policies—Filling in the 'Functional Specialty' Gaps," at pages 136–140.

[50] See the contribution of Meghan Allerton, in particular her references to Stewart Brand at note 6 on page 210.

[51] The need to link is organic, "living human bodies, linked in charity, to the joyful, courageous, wholehearted, yet intelligently controlled performance of the tasks set by a world order in which the problem of evil is not suppressed but transcended." *Insight*, *CWL* 3, 745. See also Philip McShane, *Lonergan Gatherings* 16, "(Reviewing)³ 'living human bodies linked in charity,' in the Context of the 2015 Lonergan Philosophical Society Gathering, October 8-11, ACPA, in Boston," available at: http://www.philipmcshane.org/lonergan-gatherings.

[52] It is humbling to ask in a fully serious way, "What functional specialty am I attempting?" An easier start might be to put the question to your favorite authors, as Bruce Anderson does regarding elements of dialectics, foundations, and/or policies in papers by Marin and Hutchinson. See "Haphazard Dialectics, Foundations, and Policy" at pages 117–120.

[53] See "Lonergan's Work as an Organon for Our Time," chapter one in Frederick Crowe, *The Lonergan Enterprise* (Cambridge: Cowley Publications, 1980), 1–42. "A way is traveled and left behind, but an organon functions today, and remains to function similarly tomorrow. There are many ways, some of them optional and all of them derivative from a

human growth and flourishing at all levels.[54] It represents a vitally important set of potentialities for solving profound and pressing human problems, including those stemming from what Lonergan calls the longer cycle of decline. When culture surrenders to the short-sightedness built into commonsense concern with the concrete and the immediate, the sustained and communal cultivation of "a detached intelligence that both appreciates and criticizes" becomes all the more important; its only alternative is an "ineffectualness" that will all too easily "capitulate" to blinkered policies leading to an ever-increasing social surd.[55] Functional collaboration

---

guiding intelligence, whereas an organon is unique, versatile, and adaptable to varying situations. Hence it is that one will speak of a single organon that functions in any or all of these methods." *Ibid.*, 31–32.

[54] Collaboration itself is not new, of course. We collaborate daily if we are involved in fulfilling roles and carrying out tasks in educational institutions, team sports, or even the domestic good of order when people live under the same roof. "You cook and I will clean up." Educational institutions typically divide up the roles and tasks by schools and departments, including extra-academic departments like security, food service, and athletics. Likewise function is a common experience of what soda machines do, cars do, economies do, and people do, or not, as the case may be. Thus, "my car functions" means it runs well, takes me places, makes no strange sounds, and burns not too much oil. It is easy to identify when the car is functioning well. And if it is not, we call a mechanic, the one who knows how to get it functioning again. Functional collaboration as a science is a new thing. On whom do we call, to whom do we turn, when trying to undo the misunderstanding and suffering caused by centuries of "doctrines on politics, economics, education" that cumulatively have degraded human life into something "unlivable"? *Topics in Education*, ed. Robert Doran and Frederick Crowe, vol. 10, *Collected Works of Bernard Lonergan* (Toronto: University of Toronto Press, 1993), 232. See also the following note and note 72 below.

[55] The quoted phrases appear in *Insight*, CWL 3, 262. Lonergan's search for a *novum organon* adaptable to myriad situations was long in the making. His interest in methodology stretched back to the mid-1920s and his time at Heythrop College. In an interview in *Caring about Meaning*, he recalled: "I was very much attracted by one of the degrees in the [University of] London syllabus: Methodology. I felt there was absolutely no method to the philosophy I had been taught; it wasn't going anywhere. I was interested in method." *Caring about Meaning: Patterns in the Life of Bernard Lonergan*, ed. Pierrot Lambert, Charlotte Tansey and Cathleen Going (Montreal: Thomas More Institute, 1982), 10; see also *The Question as Commitment: A Symposium*, Thomas More Institute Papers/77 (Montreal: Thomas More Institute, 1977), 10. Over fifty years later, in 1976, Lonergan would write about "the mess theology was in." "Again, *Method* was not a new idea. I was aware of the mess theology was in and considered the transposition from the question of the 'nature' of theology to the 'method' of theology to be the essential step. The work I did on *Verbum* and in *Insight* was just two stages in a program towards writing on method in theology. Indeed from 1949 to 1952 my work on *Insight* was conceived as the first part of my *Method in Theology*. But in 1952 I was told I would be teaching at the Gregorian from 1953 on, and that prompted me to publish *Insight* as a separate work." Letter to M. Lemieux, December 31, 1976. In a very

represents a solution, as well, to problems that emerge when field specialists "know more and more about less and less,"[56] live in isolated academic bubbles, and squander their efforts in endless and unproductive academic squabbles. It represents, too, a solution to problems that arise when subject specialists lack both the heuristics to collaborate with colleagues in the same department and a framework with which to cross departmental divides. Above all, it presents the possibility of thinking seriously and effectively about the human good as "a history, a concrete, cumulative process."[57] It holds the promise of enlarging the attainable human good in history, at this crucial stage of history, of caring effectively for those in need throughout all the many interdependent layers of local, regional, national, and global situations.

To those who have read Lonergan only glancingly, this may all sound rather implausible. But Lonergan's effort from the 1930s on was to discover some shared and effective way to intervene productively in the dialectic of history, some way to dispel the nightmare of what humans have made of humanity—including the immense and recurring human suffering flowing from primitively and tragically misguided economics.[58] What really did Lonergan have in mind while writing the 1969 article, "Functional Specialties"[59]—an article so far in advance of its time that it has not yet really been read, even by many of his most ardent followers? It may be hard to say, exactly, but he did link his proposed method to "actuating the potentialities" of Christian living "and taking advantage of the opportunities offered

---

real sense, Lonergan began writing *Method in Theology* in 1949. He began a draft introduction to *Method* with precisely that assertion. 69900DTE060, 3. Still later, in a 1980 Thomas More interview, Lonergan spoke of his difficulties teaching theology in Rome: "I taught theology for twenty-five years under circumstances that I consider absurd. And the reason why they were absurd was for lack of a method, or because of the survival of a method that should have been buried two hundred years ago." From an interview published in *Curiosity at the Center of One's Life: Statements and Questions of R. Eric O'Connor*, ed. J. Martin O'Hara (Montreal: Thomas More Institute, 1984), 408.

[56] *Method in Theology*, 125.

[57] *Topics in Education, CWL* 10, 33.

[58] This is most evident from his manuscripts on history, written in the mid- to late-1930s, and from his economics manuscripts, written in 1942 and 1944. But that project and promise stretches over his entire career as a thinker. How else is one to interpret the strange and haunting passages in *Insight* concerning "Cosmopolis"? And, of course, there is the evidence of the 1957 lectures on existentialism. See, e.g., *CWL* 18, 210–11 (describing "resolute and effective intervention in the dialectic" of history); *ibid.*, 306–08 (describing the conditions that must be met if "there is to be a resolute and effective intervention in this historical process"). Finally, there is a reason Lonergan gave his economics course at Boston College in the late 1970s and early 1980s the title, "Macroeconomics and the Dialectic of History."

[59] *Gregorianum* 50 (1969), 485–505, later published as chapter five of *Method in Theology*.

by world history."[60] That does not sound like someone writing for an audience of ineffectual intellectuals. What did Lonergan mean by writing, at the beginning of *Method*, that "some third way, then, must be found and, even though it is difficult and laborious, that price must be paid if the less successful subject [viz. theology, or human sciences generally] is not to remain a mediocrity or slip into decadence and desuetude"?[61] However one may eventually answer those questions, at least we may say that, given the larger arc of Lonergan's purposes, quibbles about naming the endeavor seem somewhat arbitrary. We may call it functional specialization or functional collaboration, "futurology" or "sequenomics," "whole earth and all living creatures science" or even "Cosmopolis." But the main point is that the names point to an unknown X.

> "One could give it an odd Greek-classical name, say, 'Cosmopolis.' Or one could give it a sort of Joycean Biblical-classical name, say, 'The Tower of Able.'[62] Yet call it what you will, unless you name it as strange and as not yet known, you are fatally likely to assimilate it to an existing and defective horizon rather than to recognize it as the radically novel and revolutionary thing Lonergan evidently thought it was. And whatever its name, it is, as Lonergan insists, somehow profoundly 'concerned with the fundamental issue of the historical process.'"[63]

A second element fostering common meaning and strategy is being open to learning on the go in the company of others on the go. The invitation to draft papers for the conference gathering was an invitation to participate in a process of learning a bit of science and then to critically reflect on the learning in order to better

---

[60] *Method in Theology*, 145. The link of those "potentialities" to historical process takes center stage on the last half page of *Topics in Education*. *CWL* 10, 257 (noting that "the possibilities of resisting the mechanisms and the determinisms that can emerge historically are heightened almost to an unlimited extent by Christianity. ... Christian hope ... is a supreme force in history.")

[61] *Method in Theology*, 4.

[62] See Lambert and McShane, *Bernard Lonergan: His Life and Leading Ideas*, 27, 163; McShane, *Sane Economics and Fusionism* (Vancouver: Axial Press, 2010), 26, 68–69, 84; McShane, *The Road To Religious Reality* (Vancouver: Axial Press, 2012), 45, n. 96, 49; McShane, *The Everlasting Joy of Being Human* (Vancouver: Axial Press, 2013), 106, 121, 124, n. 82.

[63] Patrick Brown, "Functional Collaboration and the Development of *Method in Theology*, Page 250," in *Himig Ugnayan: A Theological Journal of the Institute of Formation and Religious Studies*, vol. 16 (2016), special edition, *Reshaping Christian Openness: A Festschrift for Fr. Brendan Lovett*, 171–198, 181–82. The quoted phrase ending the block quotation may be found in *Insight*, *CWL* 3, 263.

understand what doing science in community means.[64] Entering into this repetitive, self-correcting process revealed to each author that he or she had much more to learn. In the fourth part of each essay, the authors reflect on their experience of attempting to collaborate scientifically.

A third element fostering common meaning and strategy concerns the symbols employed by authors in many of the essays. These symbols are place-holders that name an unknown, give minimal order to the mess, and inject a dose of humility about being able to name the functional specialties without seriously understanding what is named.[65] Whether it is C_{ij}[66] or W_3,[67] the symbolism reminds, cajoles, and forces the authors not to sit comfortably on the fence between commonsense eclecticism and scientific collaboration.[68] The symbols, you might even say, are a way of electrifying that fence. They are also a convenient way to name the road ahead and to avoid belittling the crucifying (or if you prefer, cruxifying) sacrament of the present millennium, the misbegotten effort to get by without the support of sufficiently cultured elders, men and women capable of effectively intervening in history, the history of institutions and locales, of barrios and banks, elders capable of healing or at least reducing the misery resulting from more than two centuries of defective "doctrines on politics, economics, education, and … ever further doctrines."[69] As Sean McNelis suggests, housing policy is flush with defective and destructive doctrines, and housing poverty is a symptom of the many sub-systems that are "out of sync." And as Michael George writes in his push to recycle "Finality Love, Marriage," without elderly C_{45} conversations, "it all goes to hell without a handcart (in a hurry)."[70]

---

[64] On "learning a bit of science," see the citation to "The Genetic Circle" from Lonergan's *Early Works on Theological Method I, CWL* 22, in Henman's essay at n. 14, Mundine's essay at n. 9, and McShane's Epilogue at n. 63.

[65] A relevant context is "Nominal and Explanatory Definition," *Insight, CWL* 3, 35–36.

[66] C_{ij} is shorthand for differentiated conversations between persons working in distinct functional specialties. C_{ij} represents communities of functional specialists, with the indices *i* and *j* running from 1 to 8, in internal communication. See note 9 in Quinn's essay.

[67] The meta-diagram W_3 symbolizes a hoped-for heuristic integration of specialists who have intussuscepted the hylemorphism of *Insight*, especially chapters 8, 15, and 16. W_3 was invented by McShane on a morning of the Concordia University Conference on Lonergan's Hermeneutics in November, 1986. See Pierrot Lambert and Philip McShane, *Bernard Lonergan: His Life and Leading Ideas*, 161; Philip McShane, *Prehumous* 2, "Metagrams and Metaphysics," available at: http://www.philipmcshane.org/prehumous; Philip McShane, *A Brief History of Tongue*, 124.

[68] Noting the discontinuity can be and sometimes is a regular element, for example, in reporting advances in physics, a point that Quinn makes in his essay. On commonsense eclecticism, see *Insight, CWL* 3, 441–45.

[69] *Topics in Education, CWL* 10, 230. See also notes 54 and 55 above and the citations there.

[70] See below, page 205.

A fourth element contributing to the gradual clarification and expression of common meaning flowed from an electronic exchange of draft essays before the conference gathering at the University of British Columbia. That allowed the authors to read and ponder each other's contributions in advance. For this reason one finds throughout this collection authors citing other essays in the collection. There are also references to personal correspondence (emails) from the period leading up to the conference.

The authors contributing to this collection should not be considered experts in functional collaboration, for that would set a very high bar indeed in these early stages of implementation. Rather, we are more like upper-echelon amateurs, at least in the root sense of the word, 'one who passionately seeks and pursues what he or she loves,' and perhaps also in the usual sense of 'one who is not yet fully proficient or professional.' We are grappling and groping for some foggy or not-so foggy idea about what precisely we are doing and how we may be contributing to the total end[71] of effectively intervening not just in the little mess that is theology, but also in the big mess of modernity and post-modernity,[72] the mess in which we live and move and have our non-being, and sometimes struggle to remember an integrity dormant in our most dangerous memories. Our initial and in some ways premature efforts are oriented toward seeding a very slight shift towards implementation, or effective meaning, efforts that are aligned with two methodological and transdisciplinary[73] "musts."[74]

These two 'musts' may sound far-fetched or outrageous to those born and raised and titled and published in one or other academic discipline or successful science. If the first four paragraphs of *Method* have been read at all, the 'musts' have been missed, avoided, passed over by students and disciples of Lonergan,[75]

---

[71] See *Method in Theology*, 137.

[72] The mess might find you speed-reading or simply ignoring the "massive sickness and need of axial humanity to re-globalize the wondrous darkness of spirited primates pacing and mating and poising under the moon and the clusters of clusters of ten billion galaxies." Philip McShane, *Cantower* XL, "Functional Foundations," 14. This essay is available at: http://www.philipmcshane.org/cantowers.

[73] By the time he write *Method in Theology*, Lonergan knew his proposal was interdisciplinary. See pages *Method in Theology*, 22–23, 132, and 366–67. See also the preface to three lectures by Lonergan on religious studies and theology; in those Lonergan explicitly states that *Method in Theology* was "conceived on interdisciplinary lines." *A Third Collection*, ed. Frederick Crowe (New York: Paulist Press, 1985), 113.

[74] "Some third way, then, *must* be found and, even though it is difficult and laborious, that price *must* be paid if the less successful subject [viz. theology and other human sciences] is not to remain a mediocrity." *Method in Theology*, 4 (emphases added).

[75] Lonerganism names a relatively small, effete club of post-systematic scholars who "may on occasion employ this or that technical term or logical technique but their whole mode of thought is just the commonsense mode" (*Method in Theology*, 304) and who object to

somehow mysteriously subject to a marvelous selective inattention. This collection of essays may be viewed on one level as an attempt to combat that selective inattention. It can be read as a not-yet-fully-mature attempt to "be attentive" to the mess and to make the two 'musts' a topic.

But there is also a larger task and challenge lurking here, one only touched upon above. It concerns the problem of living a truly human life in situations and institutions that may be riddled with the surd, deeply distorted or decadent, dehumanizing or degrading, thick with suffering or sick with despair. In hand-written notes titled "Aufgabe," dating from the *Insight* period, Lonergan remarked that "the problem of every generation is to figure out what can be done about the mess bequeathed it by the preceding generation."[76] What to do about the mess? We can take refuge in "what has been done," and thereby only "perpetuate the mess." Or we can "act as if nothing has been done," pretending we are not embedded in the flow of history, and place our hope in "utopian leaps."[77] Neither alternative is viable. Lonergan then framed the task or challenge before us with his customary precision.

> The problem of every generation is to live. The problem of living in any given generation has a range of solutions limited by α the proximately potential development of the existing constellation of circles of operations, β the proximately potential development of existing *Weltanschauungen*, philosophies, γ the proximately potential development of schemes of recurrence.[78]

Human living in any given generation carries forward both the accumulated achievements of the past and also its enduring messes. While maintaining the achievements requires constant effort, the messes seem almost effortlessly self-perpetuating. Yet if the problem of living a truly human life in our generation, and in future generations, is enormously complicated by the messes bequeathed to us by preceding generations, it is also a problem that can be met by drawing on the full range and resources of past and present achievement in order to face and change the future.

---

finding some third way. James Duffy has traced the sad story of fifty years (1965–2015) of Lonerganism's ineffective self-talk in "The Joy of Believing," in *Reshaping Christian Openness: A Festschrift for Fr. Brendan Lovett*, 201–227; see note 63 above.

[76] 49100D0E050, 6. "Aufgabe" in German means task, duty, function. According to the *Merriam-Webster Dictionary*, as a German technical term imported into English usage, it means "a task especially when assigned experimentally or as a test." http://www.merriam-webster.com/dictionary/Aufgabe. See also *Method in Theology*, 358, and note the resonance that Lonergan's repeated use there of the phrase, "the messy situation," has in this context.

[77] *Ibid.*

[78] *Ibid.*

Functional collaboration is, ultimately, a way of methodically and creatively expanding the range of solutions to the problem of living in any given generation. It introduces a new set of circles of operations that expands and deepens existing circles of operation; it develops the potentialities and revises the performance of existing *Weltanschauungen* and philosophies; it generates and refines new schemes of recurrence supporting cumulative and progressive improvements in scientific, academic, political, economic, social, religious, and cultural institutions and practices. It draws cyclically, methodically, and comprehensively on the human past in order to light the way to a better, more human, and more humane future. Seeding global collaboration is, in effect, a way of intervening resolutely and effectively in the dialectic of history. It is a way of seeding progress. It is a way of asking and answering the pressing and haunting question, "Do we really view humanity as possibly maturing—in some serious way—or as just messing along between good and evil, whatever we think they are?"[79]

Patrick Brown and James Duffy

---

[79] Philip McShane, "Foundations of Communications," below, 163, n. 12, quoting McShane, *The Everlasting Joy of Being Human* (Vancouver: Axial Publishing, 2013), 77.

Robert Henman

Part I. The Functional Specialty of Research

The first of the e-Seminars[1] organized by Philip McShane was on functional research.[2] It succeeded in illustrating the gap between the description in the chapter on research in *Method in Theology*[3] and the need for an introductory account of the activity. Functional research is to be a type of luminously-controlled restraint and thoroughness that is found occasionally in eccentrics, like the professor buried in the archives content to rummage round endlessly for a piece of a lost text, or like what is currently found in the field of physics, where expert observers are, or have to be, content with looking for anomalies in experimental setups.

But both the eccentric professor and the community of research physicists have an *a priori*, a context. There is the professor's background of a deep appreciation, and the mind of the research physicist is comfortable in what is called *The Standard Model*. The focus is such that data in its possible suggestiveness is tackled in a thorough fashion, helped by the functional restraint: there is, so to speak, nowhere else to go, or like the eccentric professor, there is no desire to go anywhere else.

In the seminar on functional research the convenient and familiar illustration to Lonergan readers was that of Fr. Boyer pointing out a text of Aquinas to Lonergan and claiming that it was worth looking into.[4] This raises the issue of a possible division of labor, if Boyer had passed on in an orderly fashion all the relevant texts. Think too of the accumulated and ordered anomalies that a good research team might come up with. But all this thinking and imagining cannot replace the data on functional research provided, for its understanding, by the act of actually doing it. So, here, I attempt to do functional research that may be significant for both Lonergan studies and for neuroscience. My first and central effort is towards

---

[1] There were 104 articles written for the seminar between January 15, 2011 and February 2, 2012, available at: http://www.sgeme.org/BlogEngine/archive.aspx.

[2] The story of the e-Seminar is contained in the *FuSe* series of essays available at: http://www.philipmcshane.org/fuse. In particular, functional research is the topic of Philip McShane's *FuSes* 0–9; these essays are slated to appear in a future issue of the *Journal of Macrodynamic Analysis*.

[3] Bernard Lonergan, *Method in Theology* (London: Darton, Longman & Todd, 1972).

[4] "Boyer reached for his copy of Thomas Aquinas' *Prima secundae*, pointed to an article that he himself had difficulty in interpreting, and suggested that Lonergan make a study of that article in itself, of its *loca parallela*, and of its historical sources." "Editors' Preface," *Grace and Freedom: Operative Grace in the Thought of St Thomas Aquinas*, ed. Frederick Crowe and Robert Doran, vol. 1, *Collected Works of Bernard Lonergan* (Toronto: University of Toronto Press, 2000), xviii (hereafter *CWL* 1).

manifesting the differentiation of functional research. Let me illustrate immediately the strategy in a way that echoes Boyer's suggestion to Lonergan. I simply point to a text, presented as indented text here. It is, quite literally, a page taken from current literature. I am not invoking a standard model. Like Boyer, I speak from a vague, taken-for-granted, shared, implicit standard model. Imagine now that you take the place of Lonergan, and pick-up a hint of the worthwhile attempt.

You tackle the text and gradually find your way to making explicit an answer to the problem in the context of the inadequate standard model of the scholastic tradition. You end up, like Lonergan, with a thesis. You now pause over the notion of making explicit the answer. What is it to make fully explicit the answer?

> Despite the difficult and controversial topic of providing an accurate definition, thinking is a core cognitive capacity and has traditionally been conceptualized into reasoning, problem solving and decision making. These are closely interconnected fields, although historically they have represented distinct perspectives on thinking. Reasoning, which in a broad description is drawing inferences from given information, can be subdivided into many special instances including relational reasoning, causal reasoning, conditional reasoning, analogical reasoning, and deductive and inductive reasoning. Problem solving has been defined as a goal-driven process of overcoming obstacles that obstruct the path to a solution.

> Thinking is a polymorphous term, as has been emphasized by Bennett and Hacker, who argued that for this very reason the term may not be amenable to fruitful scientific investigation. However, it is owing to this polymorphous nature that it may be used as a relevant conceptual term referring to all facets of higher cognitive processing. Additionally, thinking would also incorporate into its traditional nomenclature terms such as 'intuition,' 'insight,' 'spontaneous thought processes' and 'free floating thoughts.' One could consider a group of thinking operations (decision making, reasoning, problem solving) as explicit domains and another group (intuition, insight, spontaneous thought) as implicit domains in a taxonomy of thinking processes.

> Two theories about reasoning have dominated the cognitive literature: mental model and mental logic. Mental model is a semantic theory claiming that the central concept by which we perform reasoning operations relates to spatially organized mental models. Mental model would have predicted primarily right-hemisphere regions, especially parietal and occipital regions. In contrast mental logic claims that deductive reasoning is based on the application of formal deductive rules according to formal syntactic operations. Thus, one would expect that

left-sided prefrontal and temporal regions would be implicated in formal, rule-based operations. Over the last few years alternative and more integrative concepts have been formulated by dual-mechanism theories. These theories are presented in different versions, for instance intuitive versus deliberate, associative versus rule-based, formal and heuristic processes. These dual-mechanism concepts come closest to what one might consider as a general theory of thinking. Most of them would predict the presence of broadly distributed neural systems.

However, despite all these approaches, a coherent theory of thinking is lacking, as is a proper taxonomy for all the different flavours of its components. All in all, given the lack of data and knowledge about neuroscientific investigation into thinking on the one hand, and the missing coherent theory and taxonomy on the other hand, a book exclusively dedicated to the present state of the art of neuroimaging techniques for gaining insight into the process and organization of thinking seems warranted.

This is also justified by the impression that central domains of thinking have neither participated in nor benefited that much from the interaction of cognitive science.[5]

Let us follow the parallel. Lonergan tackled a problem within a muddled vague standard model starting from Boyer's selected text.[6] I suggest a text, say, the page indented above, and invite a tackling of the problem. Lonergan got as far as a published thesis, but did that make explicit the answer? Not according to the thinking represented by the quotation from the last chapter of *Method in Theology*. Indeed, not according to the perspective Lonergan developed on metaphysics in *Insight*,[7] where implementation became of the essence of the task of metaphysics. Within that perspective the answer is explicit when it hits the streets. Lonergan's solution, as expressed in his thesis and more fully in *Grace and Freedom*, so far from

---

[5] *Neural Correlates of Thinking*, ed. Eduard Kraft, Balázs Gulyás, and Ernst Pöppel (Berlin: Springer-Verlag, 2009), 6.
[6] One might wonder why Lonergan focused on and followed up on Boyer's suggestion. There is first the treatise pointed to generally as data, and secondly, Boyer's remark that he found the piece difficult to interpret. Would that statement arouse one's curiosity? Is there some form of handing on implicit in that remark? Yet it was effective. What of the various handing-ons in this volume, to each other, to a broader audience? A treatise needs to be written, relating to the analysis of belief in *Insight* chapter 20, on the general belief of what McShane calls Tower Workers.
[7] Bernard Lonergan, *Insight: A Study of Human Understanding*, ed. Frederick Crowe and Robert Doran, vol. 3, *Collected Works of Bernard Lonergan* (Toronto: University of Toronto Press, 1992) (hereafter *CWL* 3).

hitting the streets, has not hit the theological community. Indeed, his contribution from that thesis that is central to our own illustration from neuroscience has not as yet hit the Lonergan community, "sixty-three articles in a row ... all treat of the will."[8]

The problem that the mature functional collaboration of later centuries is to meet effectively is the problem of bringing the solution all the way to the lives of people. This is the extraordinary vision and optimism embedded in Lonergan's achievement. What it is going to demand is a thoroughness that is luminous and humble, "eliminating totalitarian ambitions"[9] while contributing effectively to total global progress.

But the beginnings of our reach for that effective vision must reach towards being potentially adequate. Here my focus is on beginnings in the first specialty, but it seems to me—and part one aims at making that clearer—that those beginnings are central to our taking Lonergan seriously. I mentioned above the "sixty-three articles in a row," and it is worth pausing over that mention as illustrating the types of entry point that we need to get moving. Functional research has to be undertaken right across the works of Lonergan if we are to have a lift-off from a half-century of Lonerganism. Let me add here, to the phrase, "sixty-three articles in a row," two other phrases of Lonergan, both from *Insight*: "pseudometaphysical mythmaking,"[10] and "there have to be invented appropriate symbolic images of the relevant chemical and physical processes."[11] These three are a sample of a multitude of phrases of doctrinal poises in Lonergan of which we must honestly say, "this deserves recycling."[12] As it happens, the three phrases noted here dominate Part Two of this essay.

But the first challenge to us here and now regards the phrase, "this deserves recycling." The phrase itself deserves recycling, and that strange notion taken to a further level helps us to see what we are up against if we are to rise slowly to a luminous collaborative structure, what McShane calls "A Tower of Able." He regularly uses symbols like "(discernment)³" to draw precise attention to unnoticed complexities in Lonergan's thought.[13] Here I am talking about the recycling of the

---

[8] Lonergan, *Grace and Freedom*, *CWL* 1, 94. The reference is to Thomas' *Prima Secundae*, qq. 6–17.

[9] "An Interview with Bernard Lonergan" edited by Philip McShane, in Bernard Lonergan, *A Second Collection*, ed. William Ryan and Bernard Tyrrell (Philadelphia: Westminster Press, 1974), 213. See also *Method in Theology*, 137.

[10] *Insight*, *CWL* 3, 528.

[11] *Insight*, *CWL* 3, 489.

[12] This phrase, suggested by Philip McShane, was the theme of the 2011 Halifax Lonergan Conference.

[13] On this topic, see Philip McShane, *The Redress of Poise* (available at: http://www.philipmcshane.org/website-books), the conclusion of chapter I, "The Value

recycling of the meant-phrase about recycling. My talking, and McShane's symbolization, point to a very remote luminosity in that not-yet-realized science of functional collaboration. Each member of each specialty is to be self-luminous about the entire human enterprise in history, a sharing destined to be luminous. It is useful here to quote a short piece headed by Lonergan with the title, *The Genetic Circle*.

> That circle—the systematic exigence, the critical exigence, and the methodical exigence—is also a genetic process. One lives first of all in the world of community and then learns a bit of science and then reflects, is driven towards interiority to understand precisely what one is doing in science and how it stands to one's operations in the world of community. And that genetic process does not occur once. It occurs over and over again. One gets a certain grasp of science and is led onto certain points in the world of interiority. One finds that one has not got hold of everything, gets hold of something more, and so on. It is a process of spiraling upwards to an ever fuller view.[14]

I am not interested here in the complex manner in which Lonergan builds the various exigencies into his functional view. What I am interested in is us taking in the bit that hits us here and now as our seeding community "learns a bit of science and then reflects, is driven towards interiority."[15] The power of the new science is that it is geared to do that effectively. But again, we cannot grasp that scientifically until it gets seriously underway. So, for instance, when it has reached the level of omnidisciplinary operation the movement from the first to the second specialty is to involve a convergence, and so on up.[16] But note—here I am returning to the subject indicated by the McShane symbolism, (turn)[3]—the "is geared to" and "is to involve" are to have a new meaning in culture and science, a new meaning emerging from becoming self-luminous about the fact that one "learns a bit of science and then reflects, is driven towards interiority." The new science is to involve a new culture of presence that, e.g., was deeply absent when Lonergan made his appeal about reading in the Epilogue of *Verbum*.[17] But to give the meaning and the appeal

---

of Lonergan's Economics for Lonergan Students," available at: http://www.philipmcshane.org/website-books.

[14] Lonergan, *Early Works on Theological Method I*, ed. Robert Doran and Robert Croken, vol. 22, *Collected Works of Bernard Lonergan* (Toronto: University of Toronto Press, 2010), 140 (1962 lectures at Regis College).

[15] *Early Works on Theological Method I*, *CWL* 22, 140.

[16] This convergence is discussed and mapped by McShane in *Cantower* 8, "Slopes: An Encounter," available at: http://www.philipmcshane.org/cantowers. See page 13.

[17] Bernard Lonergan, *Verbum: Word and Idea in Aquinas*, ed. Frederick Crowe and Robert Doran, vol. 2, *Collected Works of Bernard Lonergan* (Toronto: University of Toronto Press, 1997), 222–24.

new force I would note the discomfort of us noticing now that, just now, we did not read or hear "is geared to" or "is involved in" with that newness. Are we genuinely "driven towards interiority" by our present bit of science, indeed an optimistic drive towards an effective blossoming of humanity?

Such a simple puzzling brings us to confront the "Existential Gap"[18] in which we live and in which our present discussion occurs. The gap blocks out in us a psychological presence essential to being in history. In simpler terms it leaves us out of the Standard Model that McShane envisages as the mindset of functional collaboration, a mindset which he expresses symbolically as "FS + UV + GS,"[19] and of which he talks in terms of a "Leaning Tower of Able." In the simplest of terms, we may ask ourselves and each other about our present stance, our present interest, and find honestly that we are not psychically leaning towards being an effective ethos of progress, an ethos "that is too universal to be bribed, to impalpable to be forced, too effective to be ignored."[20]

So we arrive back, hopefully, at a fresh meaning of the phrase, "this is worth recycling." It does not matter what functional specialty we wish to join. And we may join none.

I presented above a simple page out of the literature of present neurochemical psychology. Is it worth recycling with ever widening circles, circles of inner exigencies, reaching effectively and glocally in "a process of spiraling upwards to an ever fuller view,"[21] a view that is relentlessly pragmatic? Is the 'worth' question clearly dynamic in us, so that

> it ever rises above past achievement. As genetic process, it develops generic potential to its specific perfection. As dialectic, it overcomes evil both by meeting it with good and by using it to reinforce the good. But good will wills the order of the universe, and so it wills it with that order's dynamic joy and zeal.[22]

Above I recalled Lonergan's appeal for humble creative reading, and now it is worthwhile—that issue of worth again—recalling his comment on the failure of Catholic good will, a powerful paragraph which ends with "arriving on the scene a little breathlessly and a little late."[23] But now I wish to twist again to an arrival on

---

[18] Bernard Lonergan, *Phenomenology and Logic*, ed. Philip McShane, vol. 18, *Collected Works of Bernard Lonergan* (Toronto: University of Toronto Press, 2001); see the index under *Existential Gap*.

[19] See *FuSe* 10, available at: http://www.philipmcshane.org/fuse.

[20] *Insight*, CWL 3, 263. Following that statement he begins his sketching of the characteristics of Cosmopolis.

[21] *Early Works on Theological Method I*, CWL 22, 140.

[22] *Insight*, CWL 3, 722, conclusion.

[23] *Insight*, CWL 3, 755.

the scene, the scene that is our reading of that single page of text included above, arriving, thus, freshly at the seen. We have seen the words in that text about "left-sided prefrontal and temporal regions," about "broadly distributed neural systems." Whose systems and regions are we talking about, reading about, if not our own? The page is, one might say, simply a nudge about the epigraph to *Insight* that Lonergan took from Aristotle.[24]

This gives us all a peculiar nudge about our reading of *Insight* and our self-discovery. In what sense were we discovering, or missing the discovery, of our cerebral systems? "To say it all with the greatest brevity: one has not only to read *Insight* but also to discover oneself in oneself."[25]

The discomforting thing is that when we are reading Aristotle regarding image we are indeed reading about our "left-sided prefrontal and temporal regions," about "broadly distributed neural systems." And *Insight* invites us to push on towards "complete explanation."[26] The disturbing fact is that "organic, psychic, and intellectual developments are not three independent processes. They are interlocked, with the intellectual providing a higher integration of the psychic and the psychic providing a higher integration of the organic."[27] So, "there results the problem of formulating the heuristic structure of the investigation of this triply compounded development."[28]

The fact is that what seems a topic that is innovative, grounding a massively significant invasion of the field of neuroscience, is actually a basic topic of self-understanding. Like many of the topics that emerged in McShane's e-Seminar on functional research, the question of "worth recycling" in this case is merely pointing to a flaw in the reading of Lonergan since the beginning, a flaw however, which grounds a terrible, corrupt, nominalist Lonerganism that talks in non-explanatory ways of the elements of meaning as if they were characteristics of a simple-minded disembodied Platonic entity.

However, I am here merely illustrating the general problem that we are exposing to ourselves, about ourselves, in this conference.[29] But the illustrating puts the effort of my functional research in a new context. The research has much larger

---

[24] Aristotle, *De anima,* III, 7, 431b 2. "Thus, it is the forms which the faculty of thought thinks in mental images."

[25] *Method in Theology,* 260.

[26] Both the canons of explanation, from chapters 3 and 17 of *Insight,* are relevant in this context.

[27] *Insight, CWL* 3, 494.

[28] *Insight, CWL* 3, 494–95.

[29] The conference, of course, was the Vancouver Lonergan Conference held at the University of British Columbia in July of 2014, whose proceedings are collected in this volume. The conference was titled, "Functional Collaboration in the Academy: Advancing Bernard Lonergan's Central Achievement."

scope than the lifting of neuroscience to a proper attention to its data. It also is a reach to lift Lonergan studies to a possibility of a fresh honest beginning.

So let us begin as we are, assuming that you share, at some level, the standard model I am using, the one imaged in the diagram of Appendix A here and of *CWL* 18. This diagram needs to be understood within the context of the fuller heuristic provided by W₃.[30] My research clearly brings up the question of the model that is, implicitly or explicitly, dominant in present neuroscientific research. I will only say here that it is a Scotist model that has, in fact, no seriously empirical or systematic grounding, and in particular the insight-experience has no systematic place.[31] In the conclusion to this essay I offer pointers to a reversal of the Scotist influence. Future specialties and specialty work will bring forth a fuller analysis of this faulty model, as well as a fuller presentation of the elements of meaning that makes diagrammatically evident the place of insight in the four levels of logic that are involved in human thinking.[32]

The research below points to spatial correlations between cerebral zones and elements of meaning that reveal an error in methodology due to the dominant influence of reductionism.

II. The Mind's Operations and Corresponding Cerebral Regions

A. Methods and Techniques of Research in Neuroscience

First, it will be helpful to list and distinguish briefly the different methods used to obtain this data.[33] Secondly, an attempt will be made to lineup the brain regions of

---

[30] Note that W₃ points to another lacuna in present Lonergan studies. The pointing is through the symbolic use of the semicolon—";"—which draws attention to the need to come to grips with a modern version of Aristotle's hylemorphism.

[31] Here it is best just to quote note 1 of *Method in Theology*, 336: "On conceptualism, see my *Verbum: Word and Idea in Aquinas*, London: Darton, Longman & Todd, and Notre Dame: University of Notre Dame Press, 1967, Index, *s.v.*, p. 228. The key issue is whether concepts result from understanding or understanding results from concepts."

[32] Each level of insight leap normally occurs with a full formulation. So, in answering the what-question there is a leap to a conclusion, and later the logical sequence is or can be worked out. Similarly with the leap to the answer to an is-question, whether in common sense, history, or science; the case of the other levels, what-to-do and is-to-do, is best illustrated by recipes and successful cooking. The extracted page from Kraft et al. and my text (see note 4 above) point to the need for these refinements.

[33] Charles Nelson and Christopher Monk, "The Use of Event-Related Potentials in the Study of Cognitive Development," in *Handbook of Developmental Cognitive Neuroscience*, ed. Charles Nelson and Monica Luciana (Cambridge, MA: MIT Press, 2001), 125–136, at 133 ("this method would seem to represent an ideal tool with which to explore a variety of aspects of brain-cognitive relations.")

activity corresponding with mental acts.[34] There are various types of data provided by the different methods of *mind mapping*. Six methods of gathering data are commonly used in brain chemistry research: (1) Electroencephalography techniques (EEG), (2) MRI scans, (3) fMRI scans, (4) Near-infrared Spectroscopy (NIRS), (5) Positron Emission Tomography (PET), and (6) Magnetoencephalography (MEG). I will begin with a grouping of the six methods which reveal a common characteristic.

These six methods of research and experimentation are used by neuroscientists, neurocognitive scientists, and neuropsychologists. The neuroscientist is attempting to determine the function of the various brain regions, how they function individually and how the various regions function often in unison. There are more than 82 known subregions within the 9 major regions of the brain and the interaction between these sub-zones corresponding to the acts and operations of consciousness is still largely unmapped and even less understood as to their function and interfunctioning. (See Appendix B.) There are also interactions between both hemispheres of the brain and between the human brain and the reptilian and mammalian brains which are the evolutionary precursors still part of the present human brain.

The neurocognitive scientist is attempting to form a theory of thinking. The neuropsychologist is attempting to determine the relationship between the functioning of the brain and human behavior as well as searching for the breakdowns that are responsible for diseases such as schizophrenia, autism, and Down's syndrome. All of these sciences attempt also to determine whether such brain functions or malfunctions are genetically determined, environmental, or some combination of both. A fourth science of intentionality analysis[35] is added for this discussion to aid in the development and collaboration of all three neurosciences. (See Appendix A.)

The six methods of experimentation listed above are designed to obtain data that reveal corresponding activity between the brain and the operations of the mind. Correspondence is established empirically by measuring the simultaneous occurrence [36] of the mental act and the brain activity. Verification of these

---

[34] John C. Eccles, *The Understanding of the Brain* (2nd ed)(New York: McGraw-Hill, 1977), 197 ("Brain actions give experiences and these experiences can result in thoughts that lead to a disposition to do something and so to the operation of free will—of thought taking expression in action.")

[35] Lonergan, *Insight, CWL* 3. See this text for a detailed analysis of the intentional acts of the mind.

[36] The time lapse of the electro-chemical synaptic event following the mental operation has been determined to be approximately 100 milliseconds.

simultaneous events is achieved by the repetition of the experiments.[37] The data from the first four methods consist of graphs and images that signal the occurrence of chemical-electrical changes during brain activity. PET and MEG research are designed to record changes in chemistry and magnetic fields occurring in the brain.[38]

What follows is a listing of the data from various findings to date in neuroscience, neuropsychology, and neurocognitive science. Researchers in these three neurosciences respect the present stage of their work as in its infancy in terms of the relationships between biology, psychology, and cognition.[39] *The integration of different imaging technologies is only in its infancy. Scientists have yet to fully explore interrelationships among different neurodiagnostic procedures.*[40]

This part of the paper will draw upon the findings of present neuroscience, neurocognitive science, and neuropsychology and correlate these findings with the data on the operations of the human mind. This work, of listing these correlates, provides the data that has the potential for a reorientation of neuroscience as well as expressing the differentiation of consciousness that is functional research. I add at this time that, without a prior theory of the mind's operations, neuroscientists and neuropsychologists are trapped in a broad conceptualism that restricts development in their own fields. That problematic can be resolved through the work of cyclic collaboration within the context of the functional specialties.

B. Sense Data and Brain Activity

Diagram Fig. 1 below depicts the various sensory experiences and the corresponding brain regions. The following is a listing of the correlations between the corresponding parts of the brain to the different senses.

Sense	Brain Region
Vision	Occipital and parietal lobes and cerebellum
Hearing	Temporal lobe
Touch	Outer band of cerebrum cortex

---

[37] Electrode type experiments have been ongoing for over 50 years. MRI and fMRI scanning techniques are more recent over the past two decades.
[38] PET is designed to measure oxygen and glucose changes, and MEG records changes in magnetic fields.
[39] Susan Carey, "Bridging the Gap between Cognition and Developmental Neuroscience: The Example of Number Representation," in *Handbook of Developmental Cognitive Neuroscience*, ed. Charles Nelson and Monica Luciana (Cambridge, MA: MIT Press, 2001), 415–432, at 417. Carey expresses the need for interdisciplinary research in order to work out more fully the mind/brain operations and relations.
[40] Eric Zillmer, Mary Spiers, and William Culbertson, *Principles of Neuropsychology* (Belmont, CA: Wadsworth, 2001), 222.

| Taste | Frontal lobe |
| Smell | Frontal lobe |

As depicted in the list and shown in the diagram, more than one brain region reveals synaptic activity during particular sense experience such as vision and touch. Vision activates regions related to spatiality, coordination, shape, and color. Touch activates different outer regions of the cerebrum cortex depending on which part of the body is experiencing touching.

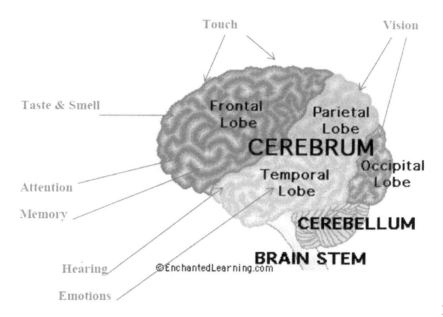

Fig. 1

C. Emotions and Feelings

Research has shown that emotions reveal synaptic activity in the limbic system which is made up of the amygdala, the caudate nucleus, the hippocampus, the hypothalamus, the putamen, and the thalamus. Each region of the limbic system performs different functions, and the frontal lobes also show activity during the experience of emotions.[41]

---

[41] Rita Carter, *Mapping the Mind* (Berkley: University of California Press, 1998), 40, 129, 162. See also Candace Pert, *Molecules of Emotion: The Science behind Mind-Body Medicine* (New York: Touchstone, Simon & Schuster, 1997), 133–135.

D. Attention

When a person focuses attention on some experience, such as visual or auditory, synaptic activity[42] increases in the frontal lobe areas.[43] The event-related potential (ERP) is believed to be primarily derived from pyramidal cells in the cerebral cortex and hippocampus.[44] The electrodes measured an increase in the brain's electrical activity during attention/memory experiments and, when sufficient neurons are activated, the current can be detected and measured against a baseline recording during a non-stimulus period or rest. Researchers provided visual and auditory experiences with different variables of intensity and time while recording the synaptic activity. This particular cortical electrode research establishes the regions of the brain in which the ERP occurs and the form of synaptic activity that is the brain pattern corresponding to the conscious element of attending.

E. Memory

Memory was tested by recording the activity when repeat experiences were provided as well as new experiences. When a test subject is utilizing their memory, synaptic activity occurs, and that activity was consistently different in a series of tests of research subjects of different age range.

Memory is believed to be the activity of drawing on storage of information which is believed to be stored in sodium compounds. Research methods have revealed that when memory is functioning, the basil forebrain structures which consist of the nucleus basalis of Meynert, the medial septal nucleus, the nucleus of the diagonal band of Broca, and the substantia innominata are all active in coordination.[45] These areas are subcortical parts of the telencephalon surrounding the inferior tip of the frontal horn and interconnected with the limbic system. The temporal lobe, putamen, hippocampus, amygdala and the caudate nucleus are also activated during particular memory tasks. Again, we witness mental activity that has corresponding activity in brain chemistry. You have a relationship here between brain chemistry and mental activity.

---

[42] Synaptic activity is the process of polarization of a cell in which ion channels occur to allow an electro-chemical transmission pass to an adjacent cell. The receiving cell absorbs the electro-chemical transmission and a transformation occurs in that receiving cell. The electrodes record this activity 100 milliseconds after it occurs.

[43] Nelson and Monk, "The Use of Event-Related Potentials in the Study of Cognitive Development," in *Handbook of Developmental Cognitive Neuroscience*, chapter 9.

[44] *Ibid.*, 125.

[45] Brian Kolb and Ian Whishaw, *Fundamentals of Human Neuropsychology* (5th ed)(New York: Worth Publishers, 2003), 468–474.

F. Knowing and Planning

Knowing is comprised of the cognitive acts that seek to correctly understand a situation or any data pertinent to a situation. So, the cognitive acts of attention to the relevant data, relevant questions, insights, judgments that formulate the insights, and finally verification of those formulations are in play when the mind is in the process of knowing. Planning takes place when the mind reflects on knowledge achieved and asks: What is to be done? That question is followed by an insight, a grasp of possible options and eventually a choosing of a particular option. Rita Carter draws on research in neuroscience that has established that the caudate nucleus of the frontal lobes is where thinking, assessing, and planning have corresponding brain activity.[46] Intentionality analysis brings to this research the various distinctions of the mental acts which Carter et al. do not include within the terms thinking, assessing, and planning. (Carter, page 298–299). The following listing outlines the order of the various mental acts and their corresponding region of brain activity.

1. Wonder (**What** question): Prefrontal Cortex. Regions differ depending on whether or not the question draws on former knowledge. If the question does not draw on former knowledge the prefrontal cortex does not show any activity. A sub-zone of the prefrontal cortex does reveal activity for questions not requiring former knowledge and draws on general cognitive abilities (Carter, page 233-239). It may be that the distinction of former knowledge is between drawing on former insights, recalled data or recalled facts.

2. Insight (Understanding): Multiple zones in the Medial Prefrontal Cortex.[47]

3. Judgment (A Formulated insight): Multiple interactive zones of the Prefrontal Cortex and frontal lobes.

---

[46] Carter, *Mapping the Mind*, 92.
[47] Mark Johnson, *Developmental Cognitive Neuroscience* (Oxford: Wiley-Blackwell, 2010), 142–43. During early childhood broader regions of the brain are activated during thinking. Upon reaching adulthood less regions are activated during thinking. It would seem that regional specialization develops over time requiring less brain regions. See also Jing Luo, Gunther Knoblich, and Chongde Lin, "Neural Correlates of Insight Phenomena," in Eduard Kraft et al., eds., *Neural Correlates of Thinking* (Berlin: Springer-Verlag, 2009), 253–67; *ibid.* at 258. Different types of insight activate different brain regions depending on whether the insight is brought on due to external, internal or restructuring hints. The common element is that insight activates many regions and usually more than other cognitive acts. The various regions activated manifest an integrative function of various brain regions just prior to and during the occurrence of an insight.

4. Is question (Seeking verification of formulation): Various zones within the Prefrontal Cortex and frontal lobes.

5. What-to-do question focused on developing options: Zones within the Prefrontal Cortex and frontal lobes.

6. Creativity: The lower part of the Prefrontal cortex: the Ventromedial or subgenual cortex. (Carter, 321).

7. Formulating options: Frontal lobe zones.

8. Insight into options: Multiple zones of the Prefrontal Cortex and frontal lobes.

9. Choosing an option: Dorsolateral region of the Prefrontal Cortex.

10. Decision to implement an option: Dorsolateral region of the Prefrontal Cortex and the Orbito-frontal cortex.

11. Act-Implementing the decision: Back of the frontal lobes: SMA: Supplementary Motor Area.

G. Will

The *will* is expressed in acts which are listed below as **Acts of the Will**. I begin with a description and account of the will found in the works of Thomas Aquinas.[48] From Aquinas we have the following.

> But a thing is in our power by the will, and we learn art by the intellect. Therefore the will moves the intellect. ... A thing is said to move in two ways: Firstly, as an end; for instance, when we say that the end moves the agent. In this way the intellect moves the will, because the good understood is the object of the will, and moves it as an end. Secondly, a thing is said to move as an agent, as what alters moves what is altered, and whatever impels moves whatever is impelled. In this way the will moves the intellect ...[49]

The will is then a motion expressed in mental operations which have corresponding activity in various regions of the prefrontal cortex. It is often described as the opposite of depression. The chemical changes initiated by depression are centered in the *locus coeruleus*.[50] The hypothalamic neurons stimulate a

---

[48] Thomas Aquinas, *Summa Theologiae*, vol. 17, *Psychology of Human Acts*, trans. Thomas Gilby, O.P. (London: Blackfriars, 1970), part I–IIa, qq. 6–17.
[49] For the Latin original of this passage, see Thomas Aquinas, *Summa Theologiae*, vol. 11, *Man*, trans. Timothy Suttor (London: Blackfriars, 1970), part Ia, q. 82, art. 4, 226–28, facing pages.
[50] See Kolb and Whishaw, *Fundamentals of Human Neuropsychology*, 727–28, for a discussion of the chemistry of depression.

secretion of corticotropin-releasing hormone when stress occurs. This stimulation is regulated by norepinephrine neurons in the locus coeruleus. Depression alters this stimulation, and a characteristic of depression is the absence or weakening of the desire to know and act. It would seem that the will no longer moves the intellect when this state is occurring. This is not to imply that regions inactive during depression are activated when the acts of will are occurring. It is probable that when the acts of will are occurring that corresponding cerebral activity is widespread and interactive but is inhibited when other areas are inactive due to depression. Because current neuroscientific literature has little to say on the acts of the will, little is known about the regions activated during the exercising of the acts of the will.[51]

The following list of terms and corresponding brain-activated regions outlines the process of the will from initial to final serenity, taken from Aquinas' analysis of the will.[52] The acts of the will are intellectual in nature but not explanatory in their reach. Therefore they may activate similar regions as do the cognitive operations listed on pages 13 and 14 but not necessarily.

Acts of the Will	Cerebral Regions
1. Initial serenity	Inactive frontal lobes
2. End	Prefrontal cortex zones
3. Value Judgment	
4. Intention	*Multiple sub-zones of the Frontal & Prefrontal lobes. During Serenity these zones show little activity.*
5. Deliberation	
6. Consent	
7. Decisive plan	Prefrontal Cortex and the Orbito-frontal Cortex
8. Choice	Dorsolateral region of the prefrontal cortex
9. Command	Prefrontal cortex zones
10. Application	Supplementary motor area
11. Achievement judgment	Prefrontal cortex zones
12. Final serenity	Inactive frontal lobe

[51] Carter, *Mapping the Mind*, 331–34. Most authors make scant reference to the terms "free will." It is too often juxtaposed with depression. Any inference of cerebral zones active during acts of the will lack an empirical base at this juncture of the research.
[52] Thomas Aquinas, *Summa Theologiae*, vol. 17, *Psychology of Human Acts*, trans. Thomas Gilby, O.P. (London: Blackfriars, 1970), part I–IIa, qq. 6–17.

H. Language

Different parts of the left hemisphere are involved in language. Spoken language activates the Wernicke's area, Broca's area is activated when speech is generated, and the angular gyrus is activated during the expression of meaning. These three areas are connected within the brain,[53] enabling all these three aspects of language to operate as a unity when expressing a spoken word or phrase. Different regions are activated depending on the stage in the movement from desiring to speak, to formulating what is to be spoken, to actually speaking. These regions do function in coordination with each other at different stages of speech. Within the context of the listing of the elements of meaning above, and the inner reach to express meaning, language would generate corresponding cerebral activity in the prefrontal cortex since language is a reaching, a desire to express, and *how* to do that is a process of early language development that becomes a spontaneous expression of meaning in childhood.[54]

The various operations of the mind activate more than one region of the brain and in most cases more than one area of a particular region. During some operations areas in both the left and right hemispheres are activated and when lesions are present in one region other regions not usually activated for a particular operation will take over in an attempt to compensate for the non-functioning zone.

---

[53] Carter, *Mapping the Mind*, 226; 255.
[54] Philip McShane, *A Brief History of Tongue: From Big Bang to Colored Wholes* (Halifax: Axial Press, 1998). This text offers an account of the emergence of language and the mental operations involved.

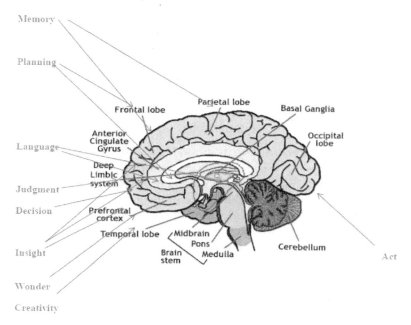

Fig. 2

I. Doubts regarding Method Outcomes

There are doubts as to whether or not a theory of mind can be achieved through scanning and imaging techniques. In the Introduction to *Neural Correlates of Thinking*, the editors, Kraft, Gulyas and Poppel highlight this problematic and Gulyas raises the same question in his article, "Functional Neuroimaging and the Logic of Conscious and Unconscious Mental Processes." He asks: "Are these techniques helping us reveal the neurobiological underpinnings of cognitive processes?"[55] What follows are comments by neuroscientists regarding their doubts about their methodology.

> What about thinking? A major theme in this book is the quest to understand thinking. The question that most reading this chapter will want to know the answer to is: "What can fMRI, or more generally, neuroimaging, contribute to our pursuit of an understanding of thinking?" Does it really help to be able to look into the brain? To borrow an analogy, can one really truly understand how computers work by

---

[55] Balázs Gulyás, "Functional Neuroimaging and the Logic of Conscious and Unconscious Mental Processes," in Kraft, et al., eds., *Neural Correlates of Thinking* (Berlin: Springer-Verlag, 2009), 141–173, at 142.

opening up a computer chassis and probing the components with a heat gun? Can identifying the when, where, and how much in the brain provide enough information so that we can begin, from this information, to derive principles of thinking? Even if we had a perfect picture at infinite spatial and temporal resolution of what was actually happening in the brain during thought, would we even then begin to understand thinking? Does it really matter what the limits of fMRI are with regard to answering questions about thinking?

It seems apparent that to truly understand the brain, a much wider context (physical and evolutionary factors) needs to be considered. Thinking itself might someday be deconstructed into simple algorithms that can be carried out within different media other than brains. Perhaps a simple model of interacting layers of neuronal networks may emerge as being able to explain thought ... It is my feeling that because thinking is a subjective process, it tends to be shrouded in mystery and potentially elevated to a status, either correctly or incorrectly, that defies understanding. ...

At the end of the day, we might be able to then say that x network, on x spatial scale, is directly related to say, theory of mind, willed action, and humor. So fMRI reveals the functions of specific processing modules. Does this really tell us anything that will help our understanding of thinking? Do we need to know what modules overlap in function or how large they are or where they are located in the brain? Does this information really matter? What spatial scale in the brain is the most critical for the understanding of thinking? While all of our tools are able to probe many different spatial scales, there are also many which have not been investigated yet. Does this matter?[56]

Horace Barlow and Rita Carter add emphases to this quandary.

... reductionism is limited because its drive is to look for explanations at lower levels in the organizational tree. ... Can we learn about the mind in the same way that we might seek to understand a machine—by taking it apart and examining its parts? Neurophysiologist Horace Barlow believes this approach can bring important insights but can never tell the full story.[57]

---

[56] Peter Bandettini, "Functional MRI Limitations and Aspirations," in *Neural Correlates of Thinking*, 15–38, at 31.
[57] Carter, *Mapping the Mind*, 18.

With the help of them (imaging techniques), can we exploit the differences between conscious and unconscious brain processes?[58]

It still remains unclear whether it is justified to assume that neural assemblies are actually the basic units of cognition.[59]

… a coherent theory of thinking is lacking … a book exclusively dedicated to … gaining insight into the process … seems warranted.[60]

The connection between neuroanatomy, neurochemistry, and neurodevelopment, and the behavioural research in cognition are rather tenuous.[61]

As always, an understanding of the mind must guide the search for its neural underpinnings.[62]

Richard Moodey offers an interesting insight into the relationship between researcher and human subjects in the following.

When working with human subjects, the neuroscientist has to ask people about their experiences in order to get information that he cannot know immediately, and relate this to his observations as an "outsider." His outsider observations are aided by ever more sophisticated apparatus, but the connections with the phenomenological accounts of the research subjects are what give fuller meaning to the external observations.[63]

Moodey's point describes the current relations operative between the human subject and the researcher. What would fill out the researcher's account? The problematic that obfuscates the settling of the issue stated in the above quotations is expressed summarily by Lonergan in the following quotation.

In this fashion, intelligence is reduced to a pattern of sensations; sensation is reduced to a neural pattern; neural patterns are reduced to chemical processes; and chemical processes to subatomic movements. The force of this reductionism, however, is proportionate to the tendency to conceive the real as a subdivision of the 'already out there now.' When that tendency is rejected, reductionism vanishes.[64]

---

[58] Gulyás, "Functional Neuroimaging," *Neural Correlates of Thinking*, 142.
[59] Michael Öllinger, "EEG and Thinking," in *Neural Correlates of Thinking*, 65–82, at 75.
[60] Kraft et al., "Introduction," *Neural Correlates of Thinking*, 6.
[61] Susan Carey, "Bridging the Gap between Cognition and Developmental Neuroscience," in *Handbook of Developmental Cognitive Neuroscience*, 415.
[62] Carey, *ibid.*, 429.
[63] Richard Moodey, email communication, on lonerganl@googlegroups.com.
[64] *Insight, CWL* 3, 282–83.

There are further questions and doubts raised in the literature about the process and method of present scanning and mapping of the brain techniques as to whether or not the outcome desired can be achieved in this manner. I pointed out in the introduction to this essay that "future specialties and specialty work will bring forth a fuller analysis of this faulty model."[65] In as much as the quotation from Lonergan expresses the procedure and error of reductionism and a possible solution for some individual, the larger task of adequate communication and implementation with both Lonergan students, Lonerganism, and the neuroscientific community[66] is only seeded here in attempting functional research.

Part III. Summary of the Research

This section of the paper will summarize what has been accomplished in Part II.

Section A listed the six techniques used in neuroscientific research in order to map the human brain.

Section B lists the cerebral correlates to activity related to the five senses.

Section C lists cerebral correlates to emotions and feelings.

Section D lists correlates to the conscious activity of attention.

Section E lists cerebral correlates to memory.

Section F lists cerebral correlates to knowing and planning.

Section G lists cerebral correlates to the will.

Section H lists cerebral correlates to language.

Section I lists doubts, appearing in the neuroscientific literature, about whether or not a theory of thinking can be achieved through techniques of scanning and imaging.

The listing of the cerebral correlates is incomplete. That said, this listing of cerebral correlates to conscious acts does manifest some patterns in terms of the more active cerebral areas during conscious activity. The frontal lobes are the most active areas. The acts of intellect correspond mostly to the prefrontal cortex and frontal lobes and all conscious acts manifest cerebral correlates in more than one area or subregion of the brain.

---

[65] Above, page 8.
[66] See my "Can Brain Scanning and Imaging Techniques Contribute To a Theory of Thinking?" *Dialogues in Philosophy, Mental, and Neuro Sciences*, vol. 6, no. 2 (December 2013) (available at http://www.crossingdialogues.com/issue22013.htm). It is an attempt to introduce cognitional theory through reflection on performance to the neuroscientific community.

The listing of the nomenclature for the five senses, memory, and the emotions and feelings are found in the neuroscientific literature. The listings of the different acts of intellect and the will are not listed in detail in the literature. These intellectual acts tend to be grouped in the neuroscientific literature under the rubric problem-solving. The acts of the will I list are grouped under the term 'will' in the literature. Familiarity with the experiences of the names of the senses and of emotions and feelings is common. The nomenclature of the acts of intellect and will as I listed them are not commonly known and are obviously absent from the literature.[67]

Sections B and C appear to be more exacting in terms of initiating and tracking of the various acts and their correlates than the other collections. Sections D, E, F, G, and H are less exact in terms of developing experiments that initiate and track the specific conscious acts involved in the human mind during memory, thinking, will, and language.

I have listed doubts at the end of Part II raised by some members of the neuroscientific community as to whether a theory of mind can be achieved through scanning and imaging techniques. These doubts became public in 2009. [68] Experimentation continues to develop techniques of scanning to provide more exacting images of brain activity, and one objective among many is to provide a theory of mind.[69]

I have, above, summarized the listings of Part II of this paper. There are patterns within these listings that suggest anomalies, and they raise questions that warrant further study. The present work of neuroscientists reveals two very different schemes of recurrence, the synaptic activity of the brain and the conscious acts and events, interrelating in the drive towards finality. These two forms of activity are believed to be related by an observation of the simultaneity of their occurrence. These two forms of activity are distinct types of data. How can this simultaneity be verified? What is the function and explanation of these apparently two simultaneous activities?

Neuroscience through its various techniques of gathering data has provided massive amounts of data from microcellular biology, biochemistry, neural chemistry, neural pathology, cerebral autopsy, scanning of synaptic activity, genetic studies, synaptogenesis, behavioral studies, brain mapping, and more. Much of the research has resulted in highly descriptive expressions of the data of the above-mentioned sciences. To date, neuroscience has been unable to answer the above questions on

---

[67] Jing Luo, Gunther Knoblich, and Chongde Lin, "Neural Correlates of Insight Phenomena," in Eduard Kraft et al., eds., *Neural Correlates of Thinking* (Berlin: Springer-Verlag, 2009), 253–67. This chapter does reflect on the act of insight as an act of intellect.
[68] *Ibid.* The specific references are noted in the footnotes of Part II.
[69] Wireless EEG have been developed which can provide more comfort during experimentation, but EEG techniques do not provide as specific locales as does fMRI. Research on combining these two techniques is ongoing. *Neural Correlates of Thinking*, 97.

an explanatory level. Genetic method, building on the descriptive work already achieved by neuroscience, would provide a method towards that outcome. This is the work of interpreters, of specialists in both functional interpretation and the many fields underlying neuroscientific research. Because the field of neuroscience involves so many subfields, interpreters who are specialists in one field would have to have some reasonable familiarity with the other related fields.

The unifying structure of the various fields would be an adequate cognitional theory, genetic method, and the functional specialist approach. That possible unity is presently blocked, or at least inhibited, on two fronts. First, the neuroscientist is presently unfamiliar with generalized empirical method[70] and genetic method as described by Lonergan in his works. Second, few Lonergan students and Lonergan scholars are specialists in neuroscience, and this is coupled with a lack of focus on functional specialization. What possible overall contribution to human advancement would an explanatory account of the relations between the human brain and human consciousness serve? A quotation from Lonergan points the way: "then theologians have to take a professional interest in the human sciences and make a positive contribution to their methodology."[71] I add philosophers to the first group and natural scientists to the latter group.

What could Lonergan have meant and how do we verify this statement? Another image might help us here: F (p_i, c_j, b_k, z_l, u_m, r_n).[72] I take this from Philip McShane's work, *Wealth of Self and Wealth of Nations*. Not unlike my listings, this is a listing of letters in a particular order. What is the pattern of meaning that these letters represent? Inasmuch as emergent probability could not bring later schemes of recurrence into existence without the functioning of the prior schemes, no science can be properly understood without understanding the scheme from which it emerged. In other words, if we desire to understand the functioning and interrelations between the brain and consciousness, we need the collaborative understanding of *p*hysics, *c*hemistry, *b*otany, *z*oology and *p*sychology if neuroscience is to move beyond description and reductionism towards a more adequate explanatory account of the relationships between the brain and the mind. This is far beyond the ability of one researcher. The functional specialist approach then becomes a necessity. Only then will both the natural and human sciences provide an adequate foundation and contribution to each of the sciences, including theology.

---

[70] See note 66 above.

[71] *Insight*, *CWL* 3, 765.

[72] Philip McShane, *Wealth of Self and Wealth of Nations: Self-Axis of the Great Ascent* (Washington, DC: University Press of America, 1975), 106. The book is available at: http://www.philipmcshane.org/published-books.

And finally, any contribution to the methodology of any science presupposes an understanding of that science.[73]

IV. Potentialities of Functional Research and Collaboration

The last section of Part Three, as an attempt to outline what is needed in the future for progress in neuroscience, is quite beyond our times. But perhaps it helps to glimpse and imagine the wider project. All I have done is make an attempt at beginning, and it was in many ways difficult for me. Focusing on gathering data tested my temperament, and I continued to be distracted by questions about the larger project. But it became evident to me eventually, that the larger project cannot be glimpsed or begun until adequate functional collaboration emerges. So, the gathering of data, searching through textbooks for correlates to conscious operations and experiences, can be a contribution to the larger project only if some individual or group takes up the task of interpreting this data and that would involve dividing up the tasks.[74] No neuroscientific group or individual to date has organized the current results of scanning and imaging of correlates to conscious operations or states in one broad source or documented place. Such a source would enable researchers to ensure the elimination of duplication of work and also act as an open-ended and ongoing development of specifications. Why has such a listing not been attempted? Such a listing would be a massive task, again requiring a fresh luminous division of labor drawing on those within many fields that comprise the entire neuroscientific effort. Such a source would reduce the possibility of duplication while providing to the enlightened an overall vibrant "image" of the ongoing work.

---

[73] See my article, "Exploring Systems and Development toward Effective Theoretic Communication," at http://roberthenman.com/articles/ Exploring%20Systems%20towards%20Effective%20Communication1.pdf, pages 1–2. See also Bernard Lonergan, "Understanding and Method" (*De Intellectu et Methodo*), trans. Michael Shields, in *Early Works on Theological Method 2*, ed. Robert Doran and H. Daniel Monsour, vol. 23, *Collected Works of Bernard Lonergan* (Toronto: University of Toronto Press, 2013), 177.

[74] Thomas Agoritsas and Gordon H. Guyatt, "Evidence-Based Medicine 20 Years On: A View from the Inside," *The Canadian Journal of Neurological Sciences*, vol. 40, no. 4, 448–49 ("the volume of research has been dramatically increasing, with now more than 2000 articles published in MEDLINE every day … Clinicians therefore need resources that filter, appraise, and synthesize the evidence").

APPENDIX A: DYNAMICS OF KNOWING AND DYNAMICS OF DOING

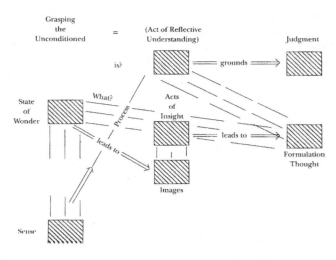

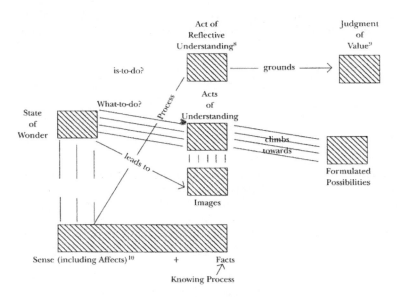

APPENDIX B: MAJOR REGIONS AND SUBREGIONS OF THE BRAIN

Frontal lobe	Superolateral	Prefontal	Superior frontal gyrus Middle frontal gyrus Inferior frontal gyrus Pars orbitalis Broca's area 　Pars opercularis 　Pars triangularis Superior frontal sulcus Inferior frontal sulcus
		Precentral	Precental gyrus Precentral sulcus
	Medial/Inferior	Prefontal	Superior frontal gyrus Medial frontal gyrus Paraterminal gyrus/Paraolfactory area Straight gyrus Orbital gyri/Orbitrofrontal cortex Ventromedial prefontal cortex Subcallosal area Olfactory sulcus Orbital sulci
		Precentral	Paracentral lobule Paracentral sulcus
	Both		Primary motor cortex Premotor cortex Supplementary motor area Frontal eye fields
Parietal lobe	Superolateral		Superior parietal lobule Inferior parietal lobule 　Supramarginal gyrus 　Angular gyrus Parietal operculum Intraparietal sulcus
	Media/inferior		Paracentral lobule Precuneus Marginal sulcus

25

	Both	Postcentral gyrus/primary somatosensory cortex Secondary somatosensory cortex Posterior parietal cortex
Occipital lobe	Superolateral	Occipital pole of cerebrum Lateral occipital gyrus Lunate sulcus Transverse occipital sulcus
	Media/inferior	Primary visual cortex Cuneus Lingual gyrus Calcarine fissure
Temporal lobe	Superolateral	Transverse temporal gyrus/Primary auditory cortex Superior temporal gyrus Wernicke's area Middle temporal gyrus–Amygdala Inferior temporal gyrus Superior temporal sulcus Inferior temporal sulcus
	Medial/inferior	Fusiform gyrus Medial temporal lobe Inferior temporal sulcus
Interlobar sulci/fissures	Superolateral	Central (frontal+parietal) Lateral (frontal+parietal+temporal) Parieto-occipital Preoccipital notch
	Medial/inferior	Media longitudinal Cingulate (frontal+cingulate) Collateral (temporal+occipital) Callosal sulcus

Limbic lobe	Parahippocampal gyrus	Anterior
		  Entorhinal cortex
		  Perirhinal cortex
		Posterior parahippocampal gyrus
		Prepyriform area

	Cingulate cortex/gyrus	Subgenual area Anterior cingulate Posterior cingulate Isthmus of cingulate gyrus: Retrosplenial cortex
		Hippocampal sulcus

	Hippocampal formation	Fimbria of hippocampus Dentate gyrus Rhinal sulcus
	Other	Supracallosal gyrus Uncus
Insular lobe	Long gyrus of insula Short gyri of insula Circular sulcus of insula	
General	Operculum Poles of cerebral hemispheres	

2. Interpreting Lonergan's Fifth Chapter of *Insight*

Terrance Quinn

I. Contexts

There are three main contexts: (1) the failure of Lonerganism regarding chapter 5 of *Insight*; (2) the confusions of present physics;[1] and (3) the functional collaboration that would help break forward in (1) and (2).

My essay tackles some aspects of the transition to (3). To whom is it addressed? To those contributing to this volume, of course, and also to those interested in Lonergan studies, even if they are not competent in physics. It does not take such competence to recognize that the functional culture changes both of the questions implicit in (1) and (2): What is philosophy? What is physics? Let us take the latter question, 'What is physics?' Normatively—in the new culture—physics is heuristically integrated into the full science of futurology in a manner that leaves it a relative functional autonomy.[2] The heuristic integration is represented by the metadiagram W_3.[3] The autonomy is evident in the manner in which the heuristic provides an open context for what is at present involved in the doing of physics, a context gently but increasingly operative. As *Method in Theology* points out, "the functional specialties of research, interpretation, and history can be applied to the data of any sphere of scholarly human studies,"[4] and so, certainly, to physics. That physics is to be lifted into the sphere of human scholarly studies is a more refined point, too complex for this short essay, yet its simplest aspect is sufficient to point to a discomforting shift in the first of the questions mentioned, "What is philosophy?": the data of philosophy always includes the concrete subjects' yearnings for adequate what-answers within human history.

All of what was stated in the previous paragraph can be communicated to common sense, although the tasks of that communication are faced adequately only

---

[1] This is not unique to physics, but is a feature of "Axial Times." See Philip McShane, *A Brief History of Tongue* (Halifax: Axial Publishing, 1998). *Axial Times* is a prolonged adolescence of humanity, a transition period that Lonergan identifies with a "second stage of meaning." See Bernard Lonergan, *Method in Theology* (London: Darton, Longman & Todd, 1973), section 3.10.

[2] I stand with McShane on the sublation of present philosophies and theologies by a cyclic-structured futurology. See his *Futurology Express* (Vancouver: Axial Publications, 2013).

[3] Pierrot Lambert and Philip McShane, *Bernard Lonergan: His Life and Leading Ideas* (Vancouver: Axial Publishing, 2010 [2nd printing 2012]), 161; Philip McShane, *Prehumous 2*, "Metagrams and Metaphysics," available at: http://www.philipmcshane.org/prehumous; Philip McShane, *A Brief History of Tongue*, 124.

[4] *Method in Theology*, 364.

in the eighth specialty. I shall return to that topic in section 4: it is an aspect of the transition to (3).

But the aspect of the transition to (3) that dominates this little essay is that aspect that involves a precision of a fresh interpretation in the cycle of functional physics and the communication of that fresh interpretation to functional historians within physics. And here, already, you can without much difficulty get a sense of a core problem of this essay, a problem that hobbles not only my effort, but the efforts of the other contributors to this volume as well. There are no functional interpreters and historians.

What I and my colleagues here are doing is, if you like, pretending that the sets of eight specialties are not empty sets. Indeed, we are pretending, even to ourselves, that there are such groups. We pretend thus in a sense that echoes the ambitions of James and Charles Stuart in 17th century England: we are "young pretenders." Indeed, the pretending relates to a new monarchy of human proceeding.[5] Unlike a Stuart pretender, however, here my pretending is to be effective. I am stepping out of line here with energetic optimism, like Luther or Einstein, with history on my side. Like Luther, I have a list of theses with complex contexts, but here, to begin to be effective I home in on two, which I call Theorem A and Theorem B.

Theorem A: "The abstract formulation, then, of the intelligibility immanent in Space and in Time is, generically, a set of invariants under transformations of reference frames, and specifically, the set verified by physicists in establishing the invariant formulation of their abstract principles and laws."[6]

Theorem B: "The concrete intelligibility of Space is that it grounds the possibility of those simultaneous multiplicities named situations. The concrete intelligibility of Time is that it grounds the possibility of successive realizations in accord with probabilities. In other words, concrete extensions and concrete durations are the field or matter or potency in which emergent probability is the immanent form or intelligibility."[7]

Part of the effectiveness is to be the style of homing in. This is best brought out by homing in on a single word of Theorem B: "situations." The word recurs in the significant context of *Method in Theology*'s reflections on "Communications": "situations are the cumulative product of previous actions."[8] The drive of this

---

[5] See Bernard Lonergan, *Phenomenology and Logic: The Boston College Lectures on Mathematical Logic and Existentialism*, ed. Philip J. McShane, vol. 18, *Collected Works of Bernard Lonergan* (Toronto: University of Toronto Press, 2001), 126–127; 130, see also the index, under *Queen*.
[6] *Insight*, CWL 3, 174.
[7] *Ibid.*, 195.
[8] *Method in Theology*, 358.

volume of essays is to lift the possibility of eight mediating situations into a normative genetics of normal-law statistics. The set of situations are to be a global topology of the situations, named C_{ij} in the full matrix of communications that is part of the shared perspective of the authors in this volume. The C_{ij} represents communities of functional specialists, with the indices i and j running from 1 to 8, in internal communication.[9] The off-diagonal elements of the matrix represent inter-functional dialogue, and my problem here is my young pretending to a reality of C_{23}. Here I note the lead given by Bob Henman in the previous essay. He operates in the first specialty seeking to draw attention to relevant data. But he adds to that effort an addressing that can be identified as an output of the eighth specialty.[10] Such an added effort relates to the shift from effeteness[11] of philosophic discourse to statistically effective discourse,[12] a lift in the concrete intelligibility of Space and Time. Here my potentially ineffective essay meshes with a like redemptive effort: there are the forthcoming books[13] that address both the community of physicists and the Lonergan community. The books seeks to invade the story of both physics and Lonerganism.

I note also that the character[14] of the seekers and the finders is a key issue here, brought out in the final fourth section.

I, we, are trying to initiate a beginning to functional collaboration. We do so in the patchy fashion illustrated by this volume to be potentially, non-Poisson, effective. Our appeal is in the togetherness of a slim heuristic FS + UV + GS tied into the shared metaword W_3 that is shared unequally among us. I go on now, in sections 2 and 3, to try to express and thus effectively hand on, in the shadow of an envisaged mature cycling, a content to both physics and philosophy that is novel. Section 2 places emphasis on discontinuity: that is a regular element, for example, in reporting advances in physics. In section 3 there will be an emphasis on continuity. That, too, is a feature of the reporting mentioned, but it is also strategic. If you like, we are inventing a new steering system, not a new vehicle.

---

[9] See Philip McShane, *A Brief History of Tongue*, 108.

[10] Bernard Lonergan, "Functional Specialties in Theology," *Gregorianum* 50 (1969): 485–504.

[11] *Method in Theology*, 99.

[12] Add *Method in Theology*, 350–51 together with *Insight*, top of page 144.

[13] Terrance Quinn, *Invitation to Generalized Empirical Method in the Sciences* (Singapore, and Hackensack, NJ: World Scientific Press); and Terrance Quinn, *The (Pre-) Dawning of Functional Collaboration in Physics* (Singapore, and Hackensack, NJ: World Scientific Press). Both books will be available in 2017.

[14] I would like to hold your interest in the concrete problems of the eighth functional specialty by quoting here Lonergan's point about character, recalling the first paragraph of the *Magna Moralia* on the fully-politicized character of theoretical thinking. "His character, in so far as it is communicative, induces in the hearer some share in the cognitive, constitutive, or effective meaning of the speaker." *Method in Theology*, 356.

It is best to get on with that effort immediately, leaving further reflection on the potential success of the two middle sections to the musings of the final section. But I would emphasize the peculiar and apparently artificial nature of these next two sections. The peculiarity and the artificialness are to vanish when the cyclic dynamic becomes a global ethos. That ethos of dynamic innovative care will then carry forward innovative corrections in a manner quite foreign to present resistance patterns to what are named paradigm shifts. Commonsense bias and staleness of roles, tasks, and institutions are to be replaced by a culture of hope effective in continually revising global care.[15]

II. The Content: Growth

A first content of my communication to functional historians, and to my fellow-writers here, is a preliminary to the significant two theorems A and B. The preliminary content is an invitation to historians, whether of philosophic history or the history of physics, that the ethos to be noted and implemented in both areas is that intimated by the doctrinal prescription:

> generalized empirical method operates on a combination of both the data of sense and the data of consciousness: it does not treat of objects without taking into account the corresponding operations of the subject; it does not treat of the subject's operations without taking into account the corresponding objects.[16]

Does this not seem a restriction on the ambition expressed in the first section? Yet, was not that ambition for effective communication and effective cycling of innovative shifts? This preliminary shift is the ground of the effectiveness, a ground absent in present history.

Still, one may wonder whether or not communicating this invitation belongs to an effort in *functional interpretation*. For instance, a somewhat similar invitation could be part of a communication C_{89}.[17] Further comment on this is given in Section 4. However, in both Sections 2A and 2B, the issue is found to be key to historical understanding.[18] Note, also, that the content of this Section 2 is preliminary description. This is not by way of "haute vulgarization," but invites future more adequate interpretations through later re-cyclings.

---

[15] The reference to the display of words on *Method in Theology*, 48, is evident. There is to be a steady increase of the dominance of the third line.

[16] Bernard Lonergan, *A Third Collection*, ed. Frederick E. Crowe (Mahwah, NJ: Paulist Press, 1985), the top lines of page 141.

[17] See third paragraph of Section 1.

[18] Implementation will only be a future achievement. See Section 4.

Below are sub-Sections 2A, 2B and 3A, 3B. Sections 2A and 2B are for theorems A and B respectively. Sections 3A and 3B are similarly organized, and are the hand-on to historians, with discussion explicitly progress-in-history-oriented.

Section 2A

"What I am trying to convey to you is more mysterious; it is entwined in the very roots of being, in the impalpable source of sensations."[19]

You will notice that Theorem A is explicit in its mention of "abstract formulation" and "invariants" that are "reached by physicists." It is, then, a theorem that partly is concerned with the "relative functional autonomy" of physics mentioned in Section 1.[20] But, as already anticipated in the first paragraphs of this Section 2, is not a core aspect of the interpretation problem immediately upon us? In the first four chapters of *Insight* leading up to chapter 5, Lonergan (compactly) expresses *his* understanding of understandings "reached by physicists." But, to what understandings can interpreters and historians appeal, luminously appeal, when reading Lonergan's Theorem A? Evidently, one needs one's own luminous understandings within physics.[21] Indeed, chapter five asks the reader to climb on from what has been attained through the four previous chapters. But, those four chapters rise on an (extraordinarily accelerated) series of exercises in self-attention in, among other things, understandings in physics and mathematics from Archimedes to Riemann, Einstein, Planck, and more.[22] And, moving into the first pages of chapter 5, the demand steepens. For instance, in section 5.1, talking about abstraction and introducing the problem of invariance, Lonergan refers to an example from chemistry: "'Pure water is H_2O.'"[23] But, certainly, "(n)o repetition of formulas can take the place of understanding."[24] Unless one is to merely repeat chemical names, there is a need here for a control of meaning in chemistry as well.

---

[19] A quotation from Paul Cézanne, recorded by J. Gasquet, *Cézanne* (La Versanne, France: Encre Marine, 2002). The original was published in 1921.

[20] See note 2.

[21] This, of course, is a normal feature of what will be generalized empirical method. And, if one is to do more than learn one's tradition and so contribute to progress, one's background needs to be up-to-date. At the present time, up-to-date physics includes quantum gauge theories. This also helps point to the fallacy of (non-functional) interpretation papers written in the old style of "Lonergan meets Balthasar."

[22] The climb is too rapid, not for Lonergan, but for what at present is the "non-existent average reader." *Insight, CWL* 3, 56. In later functional culture, content of these chapters (as well as the rest of *Insight*) will be woven into pedagogically ordered series of undergraduate and graduate courses.

[23] *Insight, CWL* 3, 164.

[24] *Method in Theology*, 351.

An important point here is that Lonergan's meaning for Theorem A remains remote to present community achievement.[25] But, what then can be said, now, of the content of Theorem A? At present, Theorem A can be taken to be a densely expressed doctrinal invitation to a (future) luminous control of meaning—in physics, and about physics. And, the theorem also points to how that can be done. For, since the theorem is explicit in its mention of understandings "reached by physicists," implicitly, is it not also methodological? Progress in understanding understandings reached by physicists is to be attained within generalized empirical method. And, what was mentioned at the beginning of Section 2 as a preliminary content and invitation, now also can be seen to belong to an interpretation of Theorem A.

A little more, though, can be obtained, even from within these preliminaries. Does the theorem refer to "all" understandings reached by "all" physicists? The larger context of *all* emergent and pre-emergent genera in physics is for discussion below, in Sections 2B and 3B. Before that, however, note that the emphasis of Theorem A evidently is on those highpoints of physical theory called "abstract principles and laws."[26] For example, there are the breakthroughs of Maxwell, Einstein, Planck, and others[27] mentioned by Lonergan, as well as Weyl, Feynman, Yang, and Mills,[28] and so on. Evidently, Theorem A focuses on those key breakthroughs in understanding which are discontinuities in the life of the physics organism. There are those physical theories by which the physics community provisionally integrally lifts, subsumes, and advances its range of capacities to perform. In other words, the understandings "reached by physicists" of Theorem A are not any and all understandings reached by physicists, but those understandings "reached (*and provisionally accepted*) by (*the community of*) physicists."[29] But, now, the admittedly vague result obtained in the previous paragraph can be sharpened somewhat. For, Theorem A now can be seen to be a doctrinal pointing to being luminous about, and within, a particular genus of development within physics.

---

[25] An exception which would prove the rule: a person who is "with" Lonergan here would be able to re-write *Insight*, or at least would be able to self-attentively "review the process from Euclidean to Riemannian geometry." *Insight, CWL* 3, 56. Although, in order to be up-to-date, such a "review" would need to include the more recent gauge theories. See the invitation of the quotation at note 62.

[26] *Insight, CWL* 3, 172.

[27] *Ibid.*, 90–91.

[28] Maxwell, Einstein, Planck, et al. are mentioned by Lonergan. I add more recent names to remind that, as we go, we will need to be including up-to-date results.

[29] I add the parenthetic phrases in italics.

Section 2B

Day of Creation[30]

Order fish, moons of the void,
To swim into the sun.
Wake birds and animals asleep
In the stony cradles of mountains.
At dusk, cast light
On a table with fruit,
A pitcher of clear water.

Theorem B is Lonergan's answer to his question: "May one not expect ... an intelligibility grasped in the totality of concrete extensions and concrete durations and, indeed, identical for all spatiotemporal viewpoints"?[31] Leading up to his answer,[32] he reminds the reader that "emergent probability exhibits generically the intelligibility immanent in world process." He goes on say: "Emergent probability is the successive realization of the possibilities of concrete situations in accord with their probabilities."[33] And, what follows is Theorem B. What, though, is any particular *situation*?

Recently, I was at a performance of *An Oriental Monsoon*, a dance troupe from Hangzhou Normal University in China.[34] The troupe's mission statement is: "We, the *Oriental Monsoon*, by the invitation of the Confucius Institutes, come to precipitate a joyful rainfall of understanding, friendship, and great love."[35] What were the *situations* the night of their performance, "capturing the spirit of classic China in the rhythm of modern music"? What is the situation now, *precipitated* in me, remembering that night?

One might wonder what these questions have to do with an interpretation of Theorem B. But, are not the Spaces and Times of performers and audience, and me now, all included in the "totality of concrete extensions and durations"? No doubt, heuristics for explaining the situations of the *Oriental Monsoon* are well beyond present community achievement.[36] But, within a preliminary W₃ perspective, among other

---

[30] Mieczyslaw Jastrun (1903–1983). Translated by Marta Zaborska and John Quinn (1956–1995). These lines are the last three stanzas of a six stanza poem, in *Dream Food, Street Editions*, vol. 3 (Toronto, 1982).

[31] *Insight, CWL* 3, 195.

[32] Here I label his answer "Theorem B."

[33] *Insight, CWL* 3, 195.

[34] The troupe was on tour, and performed at Middle Tennessee State University, February 6, 2014.

[35] *Oriental Monsoon*, Performance Program, *ibid.*

[36] See the preliminary invitation of Section 2 to "generalized empirical method."

things, there is a physics, indeed, a biophysics, of performers, performance, and audience. What, then, are the *situations* of the *physics troupe*, the physics community in history?

With the problem now in terms of *situations*, W_3 heuristics include the physics community gradually moving toward C_{ij}. But, W_3 heuristics includes not only physics, but all past, present, and future sciences, as well as all other achievements, including those of the *Oriental Monsoon*. In other words, W_3 heuristics embrace the totality, and the dynamics of the totality verifiably is a gradual movement toward an all-inclusive $(8 + 1)$-fold dynamics of situations.

Now, recall Lonergan's question, mentioned at the beginning of this section 2B: "May one not expect ... an intelligibility grasped in the totality of concrete extensions and concrete durations and, indeed, identical for all spatiotemporal viewpoints"?[37] Is not a (pre-) emergent $(8+1)$-fold all-inclusive dynamics such an intelligibility? Where Theorem A doctrinally invites toward the possibility of being luminous about, and within, a particular genus of development within physics, the invitation of Theorem B is all-inclusive. Theorem B invites us to the possibility of being luminous about, and within, the dynamic totality that is *emergent probability*. And, thanks to preliminary description within a developing W_3 heuristics,[38] emergent probability is a (pre-) emergent $(8+1)$-fold total dynamics.

III. The Handing On of the Grounded Theorems: Growing

Section 3A

Theorem A speaks of a particular kind of growth in the physics community. But, implicit in the statement of the theorem also is a call for a generalized empirical method. It was already noted in Section 2A that there is a basic problem: For, how much one lacks in understanding in contemporary physics partly is a measure of data lacking, but needed, for generalized empirical method. We must, then, pause in our tracks, or rather, in our functional trackings. Pausing so, and adjusting approach, curiously, it is the implicit yet in fact far more complex part of Theorem A upon which I briefly focus in this Section 3A. However, as is brought out below, various ongoing philosophic views of scientific method also suffer from that same basic gap in foundational development.

Within W_3 heuristics,[39] human understanding is reached through understanding experience. And it is from within that heuristics that there are questions of space, and time (or anything else). There is, however, a different

---

[37] *Insight, CWL* 3, 195; see note 33.
[38] "Our appeal is in the togetherness of a slim heuristic FS + UV + GS tied into the shared metaword W_3 that is shared unequally among us" (the second to last paragraph of Section 1).
[39] See second last paragraph of Section 1, above.

philosophic tradition in which we also find talk about experience, understanding, space, and time, but in ways that are not consistent with W₃ heuristics, and in ways that give different results. The tradition to which I refer includes Galileo (1564–1642), Descartes (1596–1650), Hobbes (1588–1679), Boyle (1627–1691), Locke (1632–1704), Berkeley (1685-1753), Kant (1724–1804), and others up to the present day.

Galileo discovered what has been called The Law of Falling Bodies. He broke through to what at the time was a new way of understanding "free-fall." Through an ingenious build-up of enquiry and experiment, he discovered a relatively simple quadratic relation for measured distances and measured times of free-fall (and that way, a constant rate of change of average speed). However, whether expressed in Latin or as ratios of imagined similar triangles, Galileo's mathematical account of free-fall evidently is nothing like seeing cannon balls drop from a tower, or musket balls roll down planks of wood. With this difference obvious, Galileo (and others) introduced a "mistaken twist" [40] in *scientific method*, involving two allegedly complementary notions: "Whereas *primary qualities*—such as figure, quantity, and motion—are genuine properties of things and are knowable by mathematics, *secondary qualities*—such as colour, odour, taste, and sound—exist only in human consciousness and are not part of the objects to which they are normally attributed."[41]

Galileo's Law is, of course, no longer considered to be explanatory of free-fall. There has been the incredible journey of physics to modern times. In that journey-climb, it was not too long before Galileo's result about free-fall was replaced by a system for all motions. That system was the one discovered by Newton (1642–1727), and invoked by the philosopher Kant (1721–1804) to justify various claims about "absolute space and absolute time." But

> if anyone were to try to bring [either] [42] the Galilean position [or Kant's ideas about absolute space and absolute time] into line with that canon [of parsimony[43]] ... [they] would have to settle an account with Einstein, who ... made various proposals regarding the space-time of physics and

---

[40] *Insight*, *CWL* 3, 107.

[41] *Encyclopedia Britannica*, "Epistemology and Modern Science," http://www.britannica.com/topic/epistemology/The-history-of-epistemology.

[42] All braces in the quoted material are my interpolations.

[43] See *Insight*, *CWL* 3, chapter three, "The Canons of Empirical Method," section 3.4, "The Canon of Parsimony," 102–107. "There is a canon of parsimony, for the empirical investigator may add to the data of experience only the laws verified in the data; in other words, he is not free to form hypotheses in the style of Descartes' vortices; but he must content himself with the laws and systems of laws exemplified by Newton's theory of universal gravitation and characterized generally by their verifiability." *Insight*, *CWL* 3, 93.

has grounds for supposing his line of thought verifiable and, to some extent, verified.[44]

And, today, the same basic challenge to all philosophic thought about space and time remains, pulling in empirical results of contemporary large-scale cosmology and Standard Model particle physics.[45]

Obviously, there are problems when a tradition of philosophic views about space and time is at odds with ongoing successes of the developing sciences, not to mention commonsense description. Galileo made a discovery that, by all accounts, was difficult to reconcile with everyday experience. And as a result, up to and including present times there remain core inconsistencies between advances in scientific understanding and some of the prevailing philosophic views. Adding to the complexity of the challenge, also included are rich contributions like those of Merleau-Ponty[46] and Renaud Barbaras.[47] Both Merleau-Ponty and Barbaras wrote about space, time, perception, and reality, but without basic appeal to either mathematics or physics.

In veterinary science, interpreting a sick puppy can mean discovering a vitamin deficiency that has been undermining normal development of the canine. In the body of philosophic views about space and time, it is evident that there are ongoing and cumulative effects of a deficiency in adequate empirical method. There is, then, the possibility of progress in historical understanding that will involve taking (functional historical) stock of what, for this paper, might be called a *vita-mens* deficiency in the developing organism of philosophic views of space and time. And, within our W3 heuristics, this brings a special focus to some of the basic goals for future functional historians: (1) make progress identifying cumulative series of twists and turns in philosophic views of space and time, that occurred through the absence of adequate empirical method; (2) include counter-factual history, and identify missed opportunities for progress which still are latent; (3) identify series of ongoing

---

[44] *Insight, CWL* 3, 109.

[45] "The old philosophic opinion that extension is a real and objective primary quality cannot dispense one from the task of determining empirically the correct geometry of experienced extensions and durations." *Insight, CWL* 3, 93.

[46] See, for example, Maurice Merleau-Ponty, "Eye and Mind," trans. Carleton Dallery, in *The Primacy of Perception and Other Essays*, ed. by James Edie (Evanston: Northwestern University Press, 1964), 159–190. A revised translation of "Eye and Mind" by Michael Smith appears in Galen Johnson, ed., *The Merleau-Ponty Aesthetics Reader: Philosophy and Painting* (Evanston: Northwestern University Press, 1993), 121–149.

[47] The "fact that we only perceive one side of the things around us doesn't mean we don't perceive them as themselves." "Introduction," Renaud Barbaras, *Desire and Distance: An Introduction to a Phenomenology of Perception*, tr. Paul Milan (Stanford, CA: Stanford University Press, 2006), 14 (original: *Le désir et la distance—Introduction à une phénoménologie de la perception* (Paris: Vrin, 1999)).

philosophic meanings about space and time that may contribute to progress, and may need to be recycled; and (4) identify ongoing meanings about space and time which, for better or for worse, have been influential, but which so far, as series, have been mainly unexplained within ongoing philosophic meaning about space and time.

In the last paragraph I mentioned *future* functional historians. What, though, of present functional historians, and the present audience? Whatever focused goals and objectives future historians work out, is there not, again, that prior task, already adverted to in connection with the explicit part of Theorem A? For, the philosophic tradition about space and time both directly and indirectly appeals to the results of Galileo, Newton, and more recently, in many cases, to the work of contemporary physicists. Evidently, a lack of self-grounded understanding in physics undermines and underminds the possibility of effective interpretation of views, as well as effective historical understanding [48] of series of views, within that philosophic tradition. What, though, about ongoing influence of works in the style of Merleau-Ponty and Barbaras? Perhaps interpretation and historical understanding of such existentially oriented works will not need the same kind of luminous foundational development in physics? But, the works of Merleau-Ponty and Barbaras are brilliant in their striving toward *fulsome descriptive knowledge of reality*. And the problem of interpreting fulsome description of reality both depends on, and goes well beyond, the *bridge* that is *Insight*'s Chapter 5.[49]

Section 3B

A key insight of Section 2B is connecting W_3 heuristics with the concrete intelligibility of Space and Time. What is pointed to is that our presence and unity in and with the dynamic universe is an (at present pre-) emergent $(8 + 1)$-fold dynamics. Does not the content of Section 2B, therefore, invite progress in functional historical understanding of that same (pre-) emergence? In particular, there is an invitation for self-luminous functional historical "review"[50] of physics and mathematics, and all other disciplines, and indeed all of human history, gripping and gripped within a developing W_3 heuristics. What, for example, are (genetic-dialectic) series that have been part of the (pre-) emergence or development of the C_{ij}? What are the (genetic-dialectic) series of shifts that partly are explained by, and partly bear out the (pre-)

---

[48] See Section 4.

[49] The "analyses of empirical science … form a natural bridge over which we may advance from our examination of science to an examination of common sense." *Insight*, *CWL* 3, 163. See also *Insight*, *CWL* 3, at 93, 107–109; and the high point that is 609–610, leading to: "it will not be amiss to draw attention to the possibility of an explanatory interpretation of a non-explanatory meaning." *Insight*, *CWL* 3, 610.

[50] *Insight*, *CWL* 3, 56.

emergence of genera and species of C_{ij}? Will not results along these lines be progress in functional history?[51]

Again, however, there is that fundamental gap between present and needed foundational development. This is revealed, for example, if we try to envisage functional historical understanding of functional history of physics (or functional historical understanding of any functional specialty in physics). Within our preliminary W_3 heuristics, is it not already evident that functional history of functional history of physics (or of any other functional specialty in physics) will need to be grounded in self-luminous development in physics? Perhaps, though, the arts will be exempt? Yet, here too, will not functional historical understanding of, say, functional foundations of musical development (or functional doctrines of musical development, and so on) depend on having experience in music, not to mention up-to-date self-luminous heuristics of musical development? But, for both performers and audience, music involves the human ear and brain, and is an achievement of the whole human organism. Evidently, any adequate heuristics for musical development will not be limited to, but certainly will need to include the neuro-physics of musical development.

IV. Concluding Contexts

Over the last few centuries, there has been remarkable progress in physics. The 20th century was an extraordinary battle-climb to the present Standard Model. But results have been *mixed*.[52] In both contemporary physics and the philosophy of physics there remain ongoing difficulties and confusions. Some of these can be associated with (non-verifiable) notions of "empty space"; there are ongoing confusions about "particles within particles"; and there are apparent paradoxes resulting when probabilities are thought to be explanatory of particular events or individual occurrences.[53] Or, from another direction, there are features of Maxwell's theory that may yet turn out to be significant in fresh ways. There is the work of Carver Mead on "Collective Electrodynamics"[54] that may yet bear fruit within a future Standard Model. There are ongoing experimental and philosophic results about Aharonov-Bohm-type effects. And, scholars in the philosophy of physics grapple with the intricacies of developing quantum gauge theories.

---

[51] There is "the problem of general history, which is the real catch." Bernard Lonergan, *Topics in Education: The Cincinnati Lectures of 1959 on the Philosophy of Education*, ed. Robert Doran and Frederick Crowe, vol. 10, *Collected Works of Bernard Lonergan* (Toronto: University of Toronto Press, 1993), 236.

[52] *Insight*, CWL 3, 514.

[53] See note 16.

[54] Carver A. Mead, *Collective Electrodynamics: Quantum Foundations of Electromagnetism* (Cambridge: MIT Press, 2002).

With this mix of pressing and open problems in meanings and ongoing meanings, my effort of passing on the two key theorems of chapter 5 of *Insight* may well seem flawed. Might it not have been better to invite readers to self-attentively enter into the details of one or more of these problems? Yet my more elementary passing on of the two theorems seems to me to fit the bill of present conditions of history, functional or not, in all areas. Various previous efforts of mine to communicate effectively "content-to-be-cycled" were altogether more elaborate and paradoxically quite flawed. What, I asked myself—and is not this metaphorical asking the ethos of the generalized empirical method and the new culture?—am I doing? Lonergan's chapter 5 of *Insight* needs to be cycled and recycled. What value, then, these earlier follies of mine, follies of summary expression? Already Lonergan's chapter is a brutal summary, and its successful reading a consequence of following up Descartes' advice given in the first paragraph of Lonergan's first chapter. At this stage, then, of our struggle for a beginning to functional collaboration, it seemed best to cycle myself back and initiate the cycling of all of my potential readers back toward that more basic task.

The basic task now appears to be one of creative re-reading of *Insight*, in the light of Lonergan's later doctrine. And I draw attention to that task for future functional historians, as I do for my colleagues in this volume. Cycling back to the more basic task also adds to Henman's pointing towards the canons of explanation,[55] by making more explicit the lift to concomitant, sentence by sentence, attention of the thinking and writing subject to what she or he is at between full stops. In our present shabby culture of what I might call, in pleasant oxymoron, truncated self-attention, this is the function of the structure of these essays that McShane suggested before we began the venture of this volume: two contextualizing parts within which there would be the functional effort. In a later mature culture of cyclic global care, the bracketing of sections 2 and 3 by two contextualizing sections will be seen as needed in the face of "the monster that has stood forth in our day"[56] before classes and in essays. But that later culture is only a dream at present, a dream of sentence–by-sentence linguistic feedback in a world of HOW-language[57] that even walks the streets beyond our schooling.

How much later is culture going to be? I suspect that it will be resisted by academic circles right through this century, though common sense and current song may join the longing for self-luminosity.[58]

---

[55] See *Insight, CWL* 3, chapters 3 and 17.

[56] *Method in Theology*, 40.

[57] "And might we find that HOW—Home Of Wonder—language?" Philip McShane, *Posthumous* 3, "A Note on Inside," 8, note 21, available at: http://www.philipmcshane.org/posthumous.

[58] Philip McShane illustrates this in his final chapter of *Method in Theology: Revisions and Implementations*, by weaving into the thirteen sections of the text the 13 songs of Sinead

But when that culture begins to emerge, its emergence will be in tandem with a luminous appreciation of how the content of the two theorems A and B flex round the cycle differently yet supplementarily. Normative invariants of the space and time of humans are to be the "cumulative and progressive results" of creative recycling. But these normativities must be discovered in this village or that, and the cyclic structure will fall in with the emergent probability lurking in Theorem B. That falling in—or rather spinning on—will aim at a normal law statistics of spreading from single village—a university village, in present culture—to a globe of villages. And such a spreading will be the potency of the blossoming of further invariants of ecohistory. So Theorem A and Theorem B will weave into fuller institutions of the genesis of roles and tasks of luminous hope and redemption.

The road there remains the road of the first paragraph of *Insight*. Sweeping paradigm shifts are not ingested by sweeping popularizations.[59] I think here, illustratively, of Richard Feynman's three pedagogical volumes on physics.[60] Only the third volume faces seriously the problem of pedagogy, and this only because the paradigm shift was altogether obscure to Feynman, to his students, to his contemporaries. We are looking at a massive change of schooling beginning with the pre-school child. A later wise culture will muse in astonishment over the obvious invariant, at home in all villages, that "when raising children you are raising questions."[61]

But historians, functional or not, cannot wait that long. The challenge historians of physics, philosophy, Lonerganism, whatever, have to face is the transposed foundational version of Lonergan's norm for the historian of mathematics:

> The history of any particular discipline is in fact the history of its development. But this development, which would be the theme of a history, is not something simple and straightforward but something which occurred in a long series of various steps, errors, detours, and corrections. Now, as one studies this movement he learns about this developmental process and so now possesses within himself an instance of that development which took place perhaps over several centuries.

O'Connor's CD, *Faith and Courage*. This book is available at:
http://www.philipmcshane.org/method-in-theology-revisions-and-implementations.
[59] See Herbert Butterfield, "Ideas of Progress and Ideas of Evolution," ch. 12 of *The Origins of Modern Science, 1300–1800* (New York: MacMillan, 1959). An online version is available at: https://archive.org/details/originsofmoderns007291mbp. In the last chapter, Butterfield acknowledges the great sweeper, Bernard le Bovier de Fontenelle. The tradition continues in most books on philosophy and history of science.
[60] *The Feynman Lectures on Physics*, edited by Robert Leighton and Matthew Sands (Boston, MA: Addison Wesley, 1964), in three volumes.
[61] Philip McShane, *Futurology Express* (Vancouver: Axial Publishing, 2013), 65.

This can happen only if the person understands both his subject and the way he learned about it. Only then will he understand which elements in the historical developmental process had to be understood before the others, which ones made for progress in understanding and which held it back, which elements really belong to the particular science and which do not, and which elements contain errors. Only then will he be able to tell at what point in the history of his subject there emerged new visions of the whole and when the first true system occurred, and when the transition took place from an earlier to a later systematic ordering; which systematization was simply an expansion of the former and which was radically new; what progressive transformation the whole subject underwent; how everything that was explained by the old systematization is now explained by the new one, along with many other things that the old one did not explain—the advances in physics, for example, by Einstein and Max Planck. Then and then alone will he be able to understand what factors favored progress, what hindered it, and why, and so forth. Clearly, therefore, the historian of any discipline has to have a thorough knowledge and understanding of the whole subject. And it is not enough that he understand it any way at all, but he must have a systematic understanding of it. For that precept, when applied to history, means that successive systems which have progressively developed over a period of time have to be understood. This systematic understanding of a development ought to make use of an analogy with the development that takes place in the mind of the investigator who learns about the subject, and this interior development within the mind of the investigator ought to parallel the historical process by which the science itself developed.[62]

One, you, you historian, must strive to read that challenge now with a fresh searching fullness. "The history of any particular discipline is in fact the history of its development." But, now one sees and is increasingly luminously seized by the story as a story of the minds of truncated searchers, often trapped in false presentations and idiot interpretations of their own doings.[63] Even Aristotle will not

[62] The quotation is from Michael Shield's 1990 translation of Lonergan's *De intellectu et methodo* (*Understanding and Method*). A more recent translation by Michael Shields of the same passage appears in "Understanding and Method," *Early Works on Theological Method 2*, ed. Robert Doran and H. Daniel Monsour, vol. 23, *Collected Works of Bernard Lonergan* (Toronto: University of Toronto Press, 2013), 175–77.
[63] Section 3 skims over this muddledness in physics but one can get a popular glimpse of this idiocy by thinking of the deductive talk of Sherlock Holmes or Poirot.

escape,[64] and certainly one must rescue Archimedes from himself: so, potential serious historians find themselves on the first page of the first chapter of *Insight*, in hot water with Archimedes.

---

[64] Place, if you like, his *Posterior Analytics* in the context of even the earliest of Lonergan's efforts, Bernard Lonergan, "The Form of Inference," in *Collection*, ed. Frederick Crowe and Robert Doran, vol. 4, *Collected Works of Bernard Lonergan* (Toronto: University of Toronto Press, 1988), 3–15.

Patrick Brown

This essay employs the four-part structure described in this volume's introduction and already familiar to readers who have struggled with the earlier essays. The first part provides a proximate context; the second attempts a limited, personal interpretation of a brief but complex passage in a 1954 letter by Lonergan; and the third asks how the interpretation can be handed forward to historians in the third functional specialty by asking what kind of heuristic perspective a functional historian might need to incarnate in order to place the interpretation, so to speak, in the pulsing flow of history.[1] Finally, I share some concluding reflections.

Unlike the other essays, this effort includes an interlude by Philip McShane. The addition of the interlude reflects a division of labor of sorts. Briefly, my attempt at interpretation concentrates only on a partial and preliminary phase of the interpretation: what can we reasonably say about the problem of expression[2] as a context for interpreting the relevant utterance, and what can we reasonably say Lonergan did *not* mean at the time by the passage in question given its proximate context. The interlude tackles the more direct and technical questions concerning the meaning of the utterance, "and proceed to the limit as in evaluating $[1 + 1/n]^{nx}$ as $\underline{n}$ approaches infinity," an utterance whose immediate context is a paragraph in the 1954 letter. It also interprets additional portions of the paragraph.

Together with the interlude, these parts cumulatively illustrate both the promise and the difficulty of our individual and collective efforts in this volume to undertake "a preliminary exploratory journey into an unfortunately neglected region"[3] of Lonergan's thought. That this region should remain so neglected nearly 45 years after the publication of *Method in Theology* is surprising, for the region

---

[1] The phrase recalls Lonergan's famous caution against abstract interpretations of his thought, found in the "Introduction" to *Insight*. Bernard Lonergan, *Insight: A Study of Human Understanding*, ed. Frederick Crowe and Robert Doran, vol. 3, *Collected Works of Bernard Lonergan* (Toronto: University of Toronto Press, 1992), 13. It also recalls, of course, the parallel phrase toward the end of the magnificent one-sentence second paragraph of chapter two of *For a New Political Economy*, ed. Philip McShane, vol. 21, *Collected Works of Bernard Lonergan* (Toronto: University of Toronto Press, 1998), 11. In the notes to follow, the *Collected Works* are abbreviated *CWL* followed by volume and page number. I reference files in the Lonergan Archives by their file number and .pdf page number. Let me also say a word about the lengthy footnotes: they would be standard in any law review article, but scholars in the other humanities may find them visually distracting. Finally, I make frequent use of explanatory parentheticals in the footnotes, a standard device in legal scholarship and a help to readers who may wish to explore the contexts of the quoted materials more fully.

[2] See, e.g., *Insight*, *CWL* 3, 576–81; 592–95; 602–03; 613.

[3] Lonergan, "Preface," *Insight*, *CWL* 3, 7.

encompasses nothing less than Lonergan's central breakthrough and the culmination of his decades-long efforts. The neglect is not only surprising but also unfortunate, because it allows Lonergan's great luminous achievement to languish in an appalling obscurity, thereby "precluding such futurible advance in knowledge"[4] of method as would otherwise promote desperately needed progress in the whole range of human sciences, with consequent precluded improvements and advances in human societies, polities, economies, cultures, and religions.

This "failure to launch," so to speak, is not only surprising and unfortunate; it is also in some ways perhaps even culpable.[5] It is, as well, more than a little disconcerting. Fred Crowe's pointed challenge and plaintive question remain unnoticed and unanswered 40 years after he framed them. "It is part of a study of Lonergan's *Method* to test it in action. When are we going to begin that implementation in theology?"[6]

1. Proximate Context

The utterance in question—reproduced below, from a letter of early May, 1954—has a prior context in Lonergan's efforts a month earlier, in April, 1954. Those efforts concerned how to settle critical and crucial issues "in the luminous fashion that will make philosophy as methodical as science," and a relevant context, too, in his insistence that the required 'luminosity' is the condition of progress on "basic and unsolved problems of theological method."[7]

---

[4] I borrow this phrase from a relevant context, Lonergan's 1943 discussion of the degrees of culpability pertaining to rationalization. "Such rationalization may involve any degree of culpability, from the maximum of a sin against the light which rejects known truth, to the minimum of precluding such futurible advance in knowledge and virtue as without even unconscious rationalization would have been achieved." "Finality, Love, Marriage," *Collection*, ed. Frederick Crowe and Robert Doran, vol. 4, *Collected Works of Bernard Lonergan* (Toronto: University of Toronto Press, 1988), 26.
[5] See Lonergan, "Dialectic of Authority," *A Third Collection*, ed. Frederick Crowe (New York: Paulist Press, 1985) 5–12, 8 (referring to the "sin of backwardness, of the cultures, the authorities, the individuals that fail to live on the level of their times.")
[6] Frederick Crowe, "Dialectic and the Ignatian *Spiritual Exercises*" (1976), in *Appropriating the Lonergan Idea*, ed. Michael Vertin (Washington, D.C.: The Catholic University of America Press, 1989), 251, n. 46.
[7] Lonergan, "Theology and Understanding" (1954), in *Collection*, ed. Frederick Crowe and Robert Doran, vol. 4, *Collected Works of Bernard Lonergan* (Toronto: University of Toronto Press, 1988), 132 (reviewing Johannes Beumer, *Theologie als Glaubensverständnis*). The article was published in vol. 35 of *Gregorianum* in October of 1954, but Lonergan had written it in the spring of 1954 (probably in April and May, since he mentions its completion in a letter to Crowe dated June 13, 1954 but not in his earlier letter to Crowe dated May 5, 1954). See the letter from Lonergan to Fred Crowe, dated June 13, 1954 ("On theological method I was pressed to review for Gregorianum J. Beumer's Theologie als Glaubensverständnis,

But must we not ask, 'luminous' *to whom*? Does there not need to be some initial, existing, functioning community in the disciplines of philosophy and theology for whom the black marks, "as methodical as science," point inwardly towards something austere, remote, demanding, something conspicuously beyond the horizon of even sophisticated common sense, something rigorously empirical and inescapably collaborative in a way that conventional philosophy and theology are not? Don't those communities need to have, or at least aspire to have, a working (as opposed to nominal) knowledge of mathematics and empirical science?[8]

And, in the context of the differentiated roles and tasks of functional collaboration, must we not also ask the question, 'luminous' *for whom*? How will "the practical insight *F*" selecting and governing flows of expression between functional specialties be transformed when the "habitual intellectual development" of "the

---

and wrote an article 29 pages plus 2 of notes.") There is another important context from that month of May. Lonergan was angling for a publisher for *Insight*, and as part of that effort, he had composed a conspectus of *Insight*, "five pages single spaced in outline of general idea," which he had sent off to Longmans "about a month ago." Letter from Lonergan to Fred Crowe, June 13, 1954. Given the time frame, Lonergan may have been composing that conspectus at the time he sent Crowe the May 5, 1954 letter. Unfortunately, it is not extant in the Lonergan or Crowe archives.

[8] Lonergan, at least, thought so, and he was not shy about saying it. Lonergan, *Method in Theology* (New York: Herder and Herder, 1972), 261–62 ("The history of mathematics, natural science, and philosophy, and, as well, *one's own personal reflective engagement in all three* are needed if both common sense and theory are to construct the scaffolding for an entry into the world of interiority" [emphasis added]); *Method in Theology*, 317 (noting "a series of fundamental changes that have come about in the last four centuries and a half. ... They involve three basic differentiations of consciousness [modern science, history, philosophy], and all three are quite beyond the horizon of ancient Greece and medieval Europe"); archive file 16230DTE070, 2 ("The scientific revolution of the seventeenth century and the historical revolution of the nineteenth constitute exigences for a remodeling of philosophy and for new methods in theology."); 89000DTE070 (transcript of question and answer session, Boston College Lonergan Workshop, June 18, 1976), 3 ("Question: In what way is an insider's knowledge of science relevant to the work of a contemporary theologian? Lonergan: Well, with respect to knowledge of science, distinguish knowledge of science in a technical fashion of one who can do science by repeating in his own development what already is known and by advancing upon that prior acquisition. That is knowledge of science in the sense in which one is the scientist; one is the mathematician; and so on. But there is also knowledge of science in an exact but schematic fashion. It doesn't attempt to be technically competent at solving differential equations and proving that infinite series converge and all the rest of it ... But you can grasp at the rudiments in calculus and analytic geometry, and so on, and are able to read with profit a book like Lindsay and Margenau's *Foundations of Physics*."); see also Lonergan's notes for the question and answer session on June 18, 1976, in archive file 27920DTE070, 5.

anticipated audience"[9] that populates the relevant handed-forward-to specialty is comfortable "in luminous fashion" with generalized empirical method? Don't we have to read both the adjective "anticipated" and the noun "audience" in Lonergan's account of the genesis of expression in *Insight* in a radically new and deeper way in light of the concrete dynamics of functional collaboration?

Lonergan's eventual solution to the "basic and unsolved problems of theological method" on his mind that Spring of 1954, elegant and luminous, involves a strategic division of labor, a corresponding structure of specialized groups, tasks, procedures, and modes of expression, and a resulting differentiation of distinct types of anticipated audiences.[10] An intimation of the challenges and difficulties involved in implementing that solution appears in the interlude and section 3 below.

Both the proximate chronological context for the paragraph in question, and the larger and more remote context concerning luminous method emerging in history, shed some light on the meaning of Lonergan's utterance. The relevant expression may be found in the first two sentences of a paragraph in a letter from Lonergan to Frederick Crowe dated May 4th, 1954.[11]

---

[9] The quoted phrases in this sentence may be found in *Insight*, *CWL* 3, 585 (the first paragraph of chapter 17, section 3, "The Problem of Interpretation").

[10] For a precise and helpful image, see the 8 x 8 matrix in Philip McShane, *A Brief History of Tongue: From Big Bang to Coloured Wholes* (Halifax: Axial Press, 1998), 108; McShane, *The Allure of the Compelling Genius of History: Teaching Young Humans Humanity and Hope* (Vancouver: Axial Publishing, 2015), 188; McShane, *Process: Introducing Themselves to Young (Christian) Minders* (Halifax: Mt. St. Vincent University Press, 1990), 106 ("The matrix represents the communications network of Cosmogenetics; so C_{23} represents a communication between a specialist in interpretation and a specialist in history."). *Process* may be found at: http://www.philipmcshane.org/website-books.

[11] I have inserted an image of the passage from the actual letter in the body of the text on the following page. Here is a version that may be easier to read: "The Method of Theology is coming into perspective. For the Trinity: Imago Dei in homine and proceed to the limit as in evaluating $[1 + 1/n]^{nx}$ as $\underline{n}$ approaches infinity. For the rest: ordo universi. From the viewpoint of theology, it is a manifold of unities developing in relation to one another and in relation to God, i.e., metaphysics as I conceive it but plus transcendent knowledge. From the viewpoint of religious experience, it is the same relations as lived in a development from elementary intersubjectivity (cf. Sullivan's basic concept of interpersonal relations) to intersubjectivity in Christ (cf. the endless Pauline [suv- or] sun- compounds) on the sensitive (external Church, sacraments, sacrifice, liturgy) and intellectual levels (faith, hope, charity). Religious experience : Theology : Dogma :: Potency : Form : Act."

> The Method of Theology is coming into perspective. For the Trinity: Imago Dei in homine and proceed to the limit as in evaluating $[1 + 1/n]^{nx}$ as n approaches infinity. For the rest: ordo universi. From the viewpoint of theology, it is a manfold of unities in developing in relation to one another and in relation to God, i.e., metaphysics as I conceive it but plus transcendent knowledge. From the viewpoint of religious experience, it is the same relations as lived in a development from elementary intersubjectivity (cf Sullivan' basic concept of interpersonal relations) to intersubjectivity in Christ (cf. the endless Pauline[suv- or]sun- compounds) on the sensitive (external Chruch, sacraments, sacrifice, liturgy) and intellectual levels (faith, hope, charity). *Rcy... Cp...... : .... : ... :: P....., F.., Act*

My personal context relates to my present efforts to co-edit Lonergan's letters, including the letters from Lonergan to Frederick Crowe and, of course, the May 1954 letter.[12] The editing is a more complex task and challenge than might initially appear. On an initial level it demands, of course, scrupulously accurate reproduction of a large series of ordered marks on a large series of pages in the Lonergan and Crowe archives. But don't those reproduced marks themselves demand a context, an adequate contextualization? If so, what would count as "adequate," and for whom?

It seems fair to assume that a fully adequate contextualization would need to move beyond contemporary conventional expectations—and with them, conventional expressions and even conventional scholarly procedures.[13] After all, we have Lonergan's own testimony that contemporarily available usage is never adequate to an original genius's purposes,[14] as well as his own conviction that the procedures of functional collaboration "will be no more received by the traditional

---

[12] With the permission and encouragement of the trustees of Lonergan's literary estate, Michael Shute and I are currently co-editing a volume to be titled *The Selected Letters of Bernard Lonergan*. The volume centers on the Lonergan-Crowe correspondence but will include a large selection of letters from Lonergan to other correspondents as well.

[13] It would be odd indeed if contemporary conventional expectations, expressions, and procedures *were* somehow up to the task of guiding an adequate contextualizing of the work of a truly innovative, polymathic genius such as Lonergan. "The old initiatives that through common acceptance have become inertial routines" (*CWL* 3, 559) are not normally ideally suited to the task of assimilating significantly new initiatives and breakthroughs or meeting significantly new challenges. To put it the other way around, were the inertial routines of present scholarship adequate to the task, Lonergan's labors regarding functional collaboration would have been quite superfluous. See n. 23 below.

[14] Lonergan, *Verbum: Word and Idea in Aquinas*, ed. Frederick Crowe and Robert Doran, vol. 2, *Collected Works of Bernard Lonergan* (Toronto: University of Toronto Press, 1997), 37 ("the original genius, precisely because he is original, finds all current usage inept for his purposes and succeeds remarkably if there is any possibility of grasping his meaning from his words"). See also *CWL* 3, 579.

understanding than Marx was received by the nineteenth century."[15] Might the present reader be operating from within some version of that "traditional understanding"? Might the present writer? Might we both be hobbled by presently available usage and presently available scholarly procedures and yet lacking in any serious sense of the extent to which that may be true?

"The possibility of exact expression of a philosophic position arises only long after the philosopher's death when his influence has molded the culture which is the background and vehicle of such expression."[16] And what of the culture to be generated by Lonergan's influence, the future culture flowing from the multi-generational operation of functional collaboration in all fields of intellectual and scientific endeavor? Is the fully adequate contextualization of Lonergan's letters to be the product of functional collaboration itself?

"The genius is simply the man at the level of his time, when the time is ripe for a new orientation or a sweeping reorganization."[17] Assuming that Lonergan was the genius that many take him to be, and assuming that functional collaboration is both a new orientation and a sweeping reorganization, would not any adequate contextualization for Lonergan's letters be the product of the novel orientation and reorganization he spent decades conceiving, namely, functional collaboration?

May we not, in fact, apply to functional collaboration itself the words Lonergan wrote in one of his early economic works? "Each stage of the long process is ushered in by a new idea that has to overcome the interests vested in old ideas, that has to seek realization through the risks of enterprise, and that can yield its full fruit only when adapted and modified by a thousand strokes of creative imagination."[18]

Obviously the task of editing in the contemporary context cannot rise to the fuller challenge of a contextualization that would profit from what contributors to the present volume presuppose about the dynamics of functional collaboration. Indeed, any effort that could somehow rise to such a level would lie somewhere beyond the present horizon of relatively-stunted-and-stilted expectations regarding the tasks and challenges of serious reading and adequate expression.[19] It would be different enough, and therefore probably off-putting enough, to offend the sense of

---

[15] Archival item 74600A0E070 (part two of a recording of a discussion between Lonergan and various professors from McMaster University in Hamilton, Ontario, on February 6, 1973 [4:35–5:48]).

[16] *Verbum*, CWL 2, 37.

[17] *Insight*, CWL 3, 444.

[18] *For a New Political Economy*, CWL 21, 20.

[19] On the problem and difficulty of developing a serious ethics of reading, see Patrick Brown, "Functional Collaboration and the Development of *Method in Theology*, Page 250," in *Himig Ugnayan: A Theological Journal of the Institute of Formation and Religious Studies*, vol. 16 (2016), special edition, *Reshaping Christian Openness: A Festschrift for Fr. Brendan Lovett*, 171–200, at 183–190.

normalcy of conventional readers and publishers. We tend to prefer what is familiar to what is strange, and what is familiar is not radically new initiatives, but "the old initiatives that through common acceptance have become inertial routines."[20]

For that reason, we may reasonably expect a transitional period—how long depends on many factors—during which readers and publishers will implacably demand that the "new wine" of the new thinking flowing from Lonergan's breakthroughs be compulsively corked in "the old bottles of established modes of expression."[21] This may seem an odd constraint to impose on interpretations and collaborations centering on a brilliantly novel thinker and his radically innovative method of collaboration. But it is not surprising. The inertia coefficient of the human mind is rather high,[22] and this inertia—in both its salutary and its sedentary effects—is relentlessly reinforced by linguistic habits[23] and patterns of expression.

---

[20] *Insight, CWL* 3, 559.

[21] *Insight, CWL* 3, 595.

[22] *For a New Political Economy, CWL* 21, 8.

[23] On the tension between what I am calling the salutary and the sedentary aspects of acquired habit, see *Insight, CWL* 3, 501. On that tension in the specific context of linguistic expression, see Lonergan's comment that "the whole tendency ... of present ways ... of speaking and doing is for them to remain as they are." *Insight, CWL* 3, 501. But that tendency, unresisted, raises the question of the genuineness of an individual or of a group. As Lonergan puts it in the section on "genuineness" in *Insight*, although "inertial tendencies" are to be respected "as necessary conservative forces," one must not on that account "conclude that a defective routine is to be retained because one has grown accustomed to it." *CWL* 3, 502. Unless we are extremely and rarely genuine in our devotion to the pure desire to know, the suggestion that a cherished routine is defective evokes "no vague tension ... but an unwelcome invasion of consciousness by opposed apprehensions of oneself as one concretely is and as one concretely is to be." *CWL* 3, 501–02.

Yet isn't the whole thrust of *Method in Theology* directed against defective intellectual and scholarly routines to which we have grown far too accustomed, and remain so to this day? We have, at least, some clues that Lonergan thought so. See, e.g., 58700DTEL60, 3–4 (early draft of ch. 1 of *Method* titled "The New Context")("In constructing the new context ... attention centers on changes in norms and procedures."); *ibid.,* 37 [page 47 in the header]) ("Once the new context is introduced, one may not revert to the old without confusion and fallacy. ... It would be a blunder, if not mere ill will, to relate the methods of the new context in the manner appropriate to relating sciences in the old context" [from a crossed out paragraph]). See also *CWL* 18, 304 ("People will see what they want to see, what can fit within their horizon, and they will omit the rest, or at least they will omit the significance of the rest. ... Insofar as people's horizon is limited, the situation can be as bad as you please and they still will not see in the situation its real significance.") The most direct and, in some ways, damning statement of Lonergan on this danger occurs in a draft introduction to *Method in Theology* circa 1966: "Apprehension of method may go no further than a set of fragmentary slogans; its acceptance may have no better basis than the other-

I find it significant that in 1966, on the threshold of writing *Method in Theology*, Lonergan expressed a conviction on this score, one apparently not widely shared by his followers. It was that "older modes of expression" may mask or "overlook" latent historical or contemporary discoveries, and that these discoveries are "capable of, begging to be expressed in the new modes. There are new bottles awaiting a new wine."[24]

Nor was his comment a rare aside, a stray tangent, or a passing instance of *obiter dicta*. The recognition that advances in knowledge both demand and result from parallel linguistic or expressive developments and refinements runs right through Lonergan's thought beginning at least in the 1940s.[25] As he contends in *Insight*, "the development of language fuses with the development of knowledge."[26] This "interpenetration of knowledge and expression implies a solidarity, almost a fusion, of the development of knowledge and the development of language."[27]

Moreover, developing thought and expression fuse in different manners at different stages of meaning,[28] and so the "techniques" of human meaning "vary in

---

directedness of conventional minds; and then its use will be unresourceful, inflexible, obtuse. The rules of the game will be known and obeyed but, unfortunately, they will not he understood; they will safeguard the prestige and privileges of an in-group, but prevent rather than promote the advance of science." 47500DTE060, 4 (nine pages from a version of chapter 1 of *Method*).

[24] Lonergan, in a letter to Roland LeBlanc, May 2, 1966, quoted in William Mathews, "A Biographical Perspective on Conversion and the Functional Specialties in Lonergan," METHOD: *Journal of Lonergan Studies* 16:2 (1998), 133–160, 149. This theme and its variations have a long lineage in Lonergan's thought. See, e.g., Lonergan, *Verbum: Word and Idea in Aquinas*, CWL 2, 37. This is, in some ways, a remarkably obvious insight, one to which Phil McShane has been calling attention for decades—repeatedly and, one must say, unsuccessfully. But I would recall here that insights are only obvious once they occur, and prior to that occurrence, their content is inherently obscure. It is revealing to think that the arc of this theme in Lonergan's thought stretches at least from *Verbum* to *Method*. For one instance in the later Lonergan's writings, see *Method in Theology*, 239 ("To be liberated from that blunder [i.e., that all knowing must somehow be like looking], to discover the self-transcendence proper to the human process of coming to know, is to break often *long-ingrained habits of* thought *and speech*." [emphasis added]).

[25] *Verbum*, CWL 2, 37.

[26] *Insight*, CWL 3, 578.

[27] *Insight*, CWL 3, 577. See also *Insight*, CWL 3, 578–79 (stressing "the genetic interpenetration of knowledge and language").

[28] "Development moves from initial, global, undifferentiated operations through differentiation and specialization to new and more effective integrations." 73400DTE060, 14; see also 131100DTE070, 7; *Method in Theology*, 287–88; *Insight*, CWL 3, 594. So also the expressive capacities and needs mediating the operations develop from initial, global, undifferentiated expression, through differentiated and specialized expressions, to new and more effective linguistic and expressive integrations. See n. 31 below.

the successive stages of man's historical development."[29] With each new stage of meaning, there occurs what Lonergan calls "the development of proportionate expression,"[30] and that development is heightened by "linguistic feed-back"[31]—as one would expect if the development of thought and the differentiation of expression go hand-in-hand. What is odd is that, while we generally recognize past instances of this important dynamic, for some reason we are far less willing to envision the necessity of future (and potentially radical) changes in expression and expressive technique.[32]

Yet Lonergan certainly anticipated such changes. One may initially think in this context, for example, of the subtleties in Lonergan's extended discussion of the "development of religious expression" as well as the "risks of religious expression,"[33]

---

[29] *Method in Theology*, 57.

[30] 73200DTE060, 31 (header 42).

[31] The key text on linguistic feedback occurs—or rather, should occur—in *Method in Theology*, page 92: "But these limitations recede in the measure that linguistic **feedback is achieved, that is, in the measure that linguistic** explanations and statements provide the sensible presentations for the insights that effect further developments of thought and language. Moreover, such advance for a time can occur exponentially; the more language develops, the more it can develop still more." *Method in Theology*, 92. The bolded language is taken from a transcript of Lonergan's reading at the 1969 Regis Institute of his typescript for that passage in *Method*; the bolding shows the line accidentally dropped sometime between the copy Lonergan read at the 1969 Institute and the copy submitted to the publisher. Compare 52300DTE060, 3, lines 17–19 with 59400DTEL70, 52 (header 131), lines 3–5. (May I note and stress here that Lonergan's otherwise obscure reference to 'exponential advances' leaps into a higher clarity and significance when the missing line is restored?) See also *Method*, 97 (noting developments based on "a rising tide of linguistic feed-back"); 73200DTE060, 42 (header 50) ("What had been going forward in the literary, philosophic, and scientific development, was an ever increasing use of linguistic feed-back."); 131000DTE070, 4 ("With literary and philosophic advance there arose the possibility of linguistic feedback: linguistic development itself provided the source of insights that grounded further linguistic development.").

[32] This odd oversight may have its origin in our almost inveterate tendency "to view language from the limited perspective of the synchronic slice we happen to be born into rather than from the sweeping diachronic perspective revealed by the whole history of human speaking and writing." Michael Shute, "Introduction: Art and the Third Stage of Meaning," *Journal of Macrodynamic Analysis* 6 (2011), 1–6, at 5; see also Shute, "Functional Collaboration as the Implementation of Lonergan's Method, Part 2: 'How Might We Implement Functional Collaboration?,'" *Journal of Macrodynamic Analysis* 8 (2015), 93–116, at 107 and n. 46.

[33] 131100DTE070, 5–9 (a detailed outline of a section titled "Religious Expression" in the "Religion" chapter in *Method* from 1968). See *Early Works on Theological Method I*, *CWL* 22, 553–58 and compare 131100DTE070, 5–9 (apparently a detailed outline of the section read

a discussion not included in the published version of *Method*. More tellingly, one may think in this context of Lonergan's compact characterization of "the fourth stage of meaning" in a draft fragment of *Method*.[34] What new techniques in the control of meaning, what innovative "developments of proportionate expression," will be required in order to express "an adequate apprehension of the transcendent," where "adequate" takes its meaning from a highly developed form of third-stage interiority? "The third stage results from meeting the critical and the methodical exigences, and the fourth adds *an adequate apprehension of the transcendent*. The third and fourth stages are those *that would result* inasmuch as transcendental method and its application to natural science, human science, philosophy, and theology become accepted."[35] Since each further stage of meaning demands shifts in expression proportionate and adequate to its advance, what unforeseen shifts in expression "would result" when functional collaboration and its application to the natural and human sciences, as well as philosophy and theology, "become accepted"?

My apparent digression on expression has two purposes. First, it serves my attempt below at interpreting the passage in the letter by highlighting the crucial issue of the adequacy or inadequacy of older or established modes of expression.

---

by Lonergan at *CWL* 22, 553 ff.). See also 73400DTE060 at 10–16 (apparently the draft text of the section read by Lonergan in 1968); 71900DTE060.

[34] John Dadosky asserts that "Lonergan never spoke of a fourth stage of meaning." "Is There a Fourth Stage of Meaning?," *Heythrop Journal: A Quarterly Review of Philosophy and Theology* 51 (2010), 768–780, 770; see also John Dadosky, "Midwiving the Fourth Stage of Meaning: Lonergan and Doran," in *Meaning and History in Systematic Theology: Essays in Honor of Robert Doran, SJ*, ed. John Dadosky (Milwaukee: Marquette University Press, 2009), 71–92, 72 (suggesting that "we may never know if Lonergan would approve of a fourth stage of meaning.") As it happens, these assertions are not entirely correct. It is true that none of Lonergan's published works, to my knowledge, use the phrase. But the "fourth stage of meaning" occurs twice in an important passage in a surviving fragment of a draft for the "Meaning" chapter in *Method in Theology*. See the following footnote and the text it references.

[35] 73200DTE060, 26 (header 42)(discards from a draft of the chapter on "Meaning" circa 1968)(emphasis added); *ibid.*, 27 (also header 42) ("The nature of the fourth stage will be indicated in the next chapter," that is, the then-projected chapter on religion. See 73200DTE060, 24 (header 32), lines 5–6. Let me illustrate this briefly and haltingly, if I may. Suppose a further stage of meaning, past the stage characterized by the highly developed differentiation of interiority, in which the attainment of "an adequate apprehension of the transcendent" occurs, an attainment that meets the conditions stipulated by Lonergan in that passage. What changes in grammar, in language, in gesture, in self-conception and self-reference, in techniques of linguistic feed-back, in religious expression and its risks, what changes in culture and pedagogy, worship and witness, would be proportionately required by the fullest possible realization that "the universe is in love with God" (*Insight, CWL* 3, 721), molecules and all, and that "God is not an object"? *Method in Theology*, 341–42; see also 131100DTE070, 13–14.

Second, it is an attempt to contextualize the need for a future contextualization of Lonergan's achievement, including his letters, especially the letter of May 4, 1954, and its central paragraph. It is a contextualization-in-progress, so to speak, and will be for some time. Still, when the task of editing the letters is eventually done in an acceptable contemporary sense, I would hope that this initial foray into interpreting functionally part of a single paragraph of one of the letters might help to lift the entire reading of the letters—at least for me and for others who try the experiment.

One benefit of struggling with the first two sentences of this particular paragraph is that one simply cannot glide past them, or the longer paragraph in which they appear, as though one somehow already knew what Lonergan meant, or as though Lonergan were gratuitously showing off his mathematical prowess when he could have just as easily and adequately expressed his new "perspective" on the method of theology in simple, plebeian, and accessible terms. Quite to the contrary. One is brought up short in reading those sentences precisely because—quite conspicuously—one lacks the all-too-usual comfortable illusion of a readily available meaning to the text, together with its all-too-human companion, a predictably *simpliste* reading of the text. A facile crypto-nominal reading is simply not an option here.

Functional interpretation, normatively, is anything but facile. I am attempting in this essay to interpret portions of only a few sentences. Yet like the "fresher and deeper rhythms"[36] of perception flowing from the aesthetic or artistic break from "the ready-made subject in his ready-made world,"[37] cycles of functional interpretation will someday evoke fresher and deeper rhythms of interpretation[38] over a

---

[36] See *Method in Theology*, 29; *ibid.*, 61–64.

[37] *Method in Theology*, 62.

[38] I am thinking here not only of rhythms generated by cycles of functional collaboration within a given generation but also of cumulative and progressive collaboration across generations of interpreters. See *Insight*, *CWL* 3, 610 ("The totality of documents cannot be interpreted scientifically by a single interpreter or even by a single generation of interpreters. There must be a division of labor, and the labor must be cumulative.") And, Lonergan would later add, the labor must be 'progressive' across generations. Contemporary human sciences are certainly cumulative in the sense that their printed products weigh more and more metric tons each year. But it is not at all clear that their results are cumulative in the normative sense of being increasingly the product of increasingly effective and efficient collaboration—not to mention that what you might call their future 'effectively progressive' status seems far beyond present proximate practice or even present fantasy—a possibility Lonergan did not fail to envision. See, e.g., *Insight*, *CWL* 3, 659 ("To grasp his own developing is for man to understand it, to extrapolate from his past through the present to the alternative ranges of the future. It is to extrapolate not only horizontally but also vertically, not only to future recurrences of past events, but also to future higher integrations of contemporary unsystematized manifolds.")

much wider range of Lonergan's writings—eventually fostering, one might say, a break from the ready-made Lonergan and his ready-made readers.

"But we are not there yet."[39] And I am not there yet. As the stark contrast provided by McShane's interlude reveals, my own efforts here fail, in the end, to escape the gravitational pull of *haute vulgarization* in presently defective academic culture. Still I hope that these efforts may point in some suggestive way to the future potential lifting involved in creative and precise functional interpretation. Stated otherwise, however faltering and inadequate this initial effort may be, and must be, I hope it intimates in some way the effect that ingesting and absorbing a portion of the meaning of the chosen paragraph can have on functional interpretation of the letters, as well as on the reflections to follow in functional history. Perhaps some elements in the interpretation will be worth cycling, or recycling, through further spirals of functional collaboration.

Let us briefly try the experiment, then, of generating an increment in a freshened and deepened perspective—in this case, a perspective on sequences of utterances. Think for a moment of the process of reading that moves on from this particular letter to the next. The same man who wrote the next letter also wrote this paragraph of this letter. Does that statement, does that realization, seem embarrassingly banal and obvious to you, almost unworthy of notice? That is certainly the most obvious and banal interpretation of that statement, that realization. But what if you were willing to risk moving beyond the banal and the obvious?

Does the realization not, rather, invite a fresher, deeper, and more demanding reading of the next letter? Will it not put the reader of later letters on notice of a new and more demanding standard of adequate reading, on edge, as it were, on alert for levels and depths of meaning previously missed? Moreover, if this suggestion is genuinely savored and implemented—"sensed" in a displacement of the Ignatian sense of "sensed" to the world of explanatory interiority—doesn't the whole series of letters receive, so to speak, an elevating grace, a release, an upward shift or lift, so that a re-reading would take place within a quite new context?

Next, think of the functional historian who is interested in the effect in history, or on history—starting with Crowe—of reading these letters. Put aside for a moment the depressing realization that the letters fall within a body of works that so far has had little effect on history, very little *Wirkungsgeschichte*, as the Germans might say. Or perhaps, on second thought, we should not put that aside at all. Perhaps that seemingly depressing realization is precisely the point. The possibility of resisting and eventually reversing that ineffectuality, that neglect, that gross negligence, is in some sense a hope riding on present and future functional interpretation, functional history, and functional dialectic.

---

[39] *For a New Political Economy*, CWL 21, 20.

2. Interpretation

I want now to move from the significance of its relevant, proximate context to the task and problem of interpreting the specific text. The first step is to encourage you to ask with me, within yourself, is there in fact a task or problem at all? Can't we legitimately know, with relatively little concentrated effort, the basic gist of what Lonergan was trying to say to Fed Crowe in May, 1954? Don't his words on the page, however initially obscure, provide us with enough to arrive at something like "a knowledge, a dim earnest,"[40] a kind of hermeneutic gleaning or gloaming, of what he meant, what he intended, what he wanted to communicate to Fred Crowe in that May month of 1954, in that letter, in that paragraph, in those first two sentences, regardless of the further complexities and depths that scholars may later identify?

I would say "no," and that "no" may jar your spontaneous commonsense anticipation that basic "gists" and intelligent "gleanings" can suffice to provide a "general idea" of what Lonergan meant. My no-saying also helps to define the problem, or rather, problems. The question involves not only one problem, but two distinct problems. The first I would associate with a fuller challenge of interpretation, and the second I would connect with Fred Crowe's efforts to understand why Lonergan would attempt to address him in such an apparently gnomic manner.[41] Oddly, the problem of interpreting the text within the fuller context may be less difficult than the second problem. For the second problem involves, among other things, teasing out the precise empirical components and counterparts to what Lonergan called "the variable standard of adequate expression,"[42] or, stated otherwise, the problem of communicating within, and to, a deficient context.[43] That complex problem area bristles with questions and challenges, and exploring it here would take us too far afield from the present task.

So instead I want to move in the direction of a precise interpretation of the text, in contrast to the defective anticipation of gleaning its general meaning, and to

---

[40] William Wordsworth, "The Prelude" (1799), in *The Essential Wordsworth*, selected by Seamus Heaney (Hopewell, NJ: The Ecco Press, 1988), 88. The 'words on the page' are mythical, of course. The words are not on the page at all, though one might say that the marks on the page invite us to evoke our capacities for attention, intelligence, reasonableness, and responsibility, "the interpreter's capacity to grasp meanings" that forms "the sources of interpretation" immanent in us. *Insight*, *CWL* 3, 588.
[41] In the spring of 1981, Fred Crowe sent a letter to Phil McShane asking for help in interpreting the equation in Lonergan's letter of May 5, 1954. See letter of Frederick Crowe to Philip McShane dated March 13, 1981.
[42] *Insight*, *CWL* 3, 580.
[43] I return to this topic briefly in section 3.

the difficulty on which I wish to focus.[44] That difficulty is the meaning Lonergan had for "in evaluating $[1 + 1/n]^{nx}$ as $\underline{n}$ approaches infinity." Let us raise and risk a relevant question, one perhaps oddly liable to being pre-consciously stifled in our psyches by the neural correlates of the constructive censor as dominated both by the general bias and by the mythic and defective anticipation that complex meaning can be bounded in a nutshell. The question is this: Did Lonergan have in mind precise meanings of *n* and *x* when he wrote that? Or was his typed text simply a matter of helpfully intimating the procedure of taking a limit, a sort of stately or impressive but ultimately vague way of communicating a generally accepted view of growth in theology or increase in understanding? But vague and general views, stately or otherwise, do not seem an accurate fit with Lonergan's confident declaration that the method of theology was "coming into perspective."[45]

What was "coming into perspective," and why did Lonergan select that equation to express his growing perspective on the method of theology? One can verify an interpretative hypothesis only in the mode of increasing or converging probabilities. In contrast, under some conditions, one can falsify an interpretative hypothesis with certainty.[46]

Let us begin, then, with the simpler negative task. It is possible to say with relative certainty what Lonergan's text did *not* mean. We can dispense, for example, with interpretative hypotheses that take the low road of what Lonergan had already at that point in his thinking named "the popular fallacy."[47] "If often enough the meaning of an expression is simple and obvious, why should it not always be so? Why should honest truth ever hide in the voluminous folds of a lengthy, complicated, and difficult exposition?"[48]

The meaning of an expression is simple and obvious to a reader or listener when the reader or listener already possesses the insights animating the expression. Within the realm of commonsense knowing, the animating insight may be relatively

---

[44] The rest of the complex 1954 paragraph needs interpretation, indeed, a much larger and deeper treatment than I am able to provide. James Duffy's essay touches on aspects of the rest of the paragraph, as does Phil McShane's interlude below.

[45] Nor does a vague and general view about Lonergan's vague and general views account for the other mathematical analogies Lonergan uses to illustrate precise notions concerning growth in theological understanding in history. See, e.g., *Early Works on Theological Method I*, *CWL* 22, 171–72.

[46] See "On God and Secondary Causes," *Collection*, *CWL* 4, 60.

[47] *Insight*, *CWL* 3, 581. See also Philip McShane and Pierrot Lambert, *Bernard Lonergan: His Life and Leading Ideas* (Vancouver: Axial Publishing, 2010 [second printing, 2013]), 252 (recounting the story of Lonergan being asked by a member of his community when Lonergan would talk more plainly, together with Lonergan's humorous and piquant reply: 'Your philosophy is the philosophy of the dog. If you can't chew it, or screw it, you piss on it.'")

[48] *Insight*, *CWL* 3, 581.

easy to attain. Serious theoretic or scientific insights, however, emerge only within elaborate and extensive frameworks painfully constructed over decades or centuries, and one acquires technical knowledge of the science only "by repeating in his own development what already is known and by advancing upon that prior acquisition."[49] Accordingly, theoretic or scientific insights find "adequate expression only in the abstract and recondite formulations of the sciences."[50] Evidently Lonergan was not proffering a commonsense insight to Crowe.

When a man of Lonergan's theoretic caliber, who has been hard at work for at least 15 years attempting to remedy a lack of method haunting a decadent theology,[51] says that "the Method of Theology is coming into perspective," he is not engaging

---

[49] 89000DTE070 (transcript of question and answer session, Boston College Lonergan Workshop, June 18, 1976), 3.

[50] *Insight*, *CWL* 3, 30.

[51] I am counting from the years 1939 or 1940, the years in which elements of Lonergan's methodological *tour de force* in the "Introduction" and "Form of the Development" chapters to his dissertation were drafted. See *Grace and Freedom: Operative Grace in the Thought of St Thomas Aquinas*, vol. 1, *Collected Works of Bernard Lonergan*, ed. Frederick Crowe and Robert Doran (Toronto: University of Toronto Press, 2000), 155–192. See also *ibid.*, 155 ("It remains that, though a method which solves the problem is possible, its use makes extreme demands on the reader.") Is this sentence-length-passage from the youthful, energetic, and obviously forward-leaning Lonergan, perhaps, a compact and programmatic compendium of virtually all of Lonergan's works on method, from *Grace and Freedom* through *Insight* to the discovery of functional specialization, its compact formulation in *Method in Theology*, and extending as well in subtle ways to the later writings? If that is so, what does it say about what one might call the variable standard of adequate reading? Do most scholars and students approach their reading (or re-reading) of Lonergan knowing full-well that it will make "extreme demands" on them?

My sense is that this is not a dominant, widely-shared *ethos* of Lonergan studies; certainly very few Lonergan students and scholars appear to think the method of functional specialization makes "extreme demands" on them, or that it solves a problem that has become acute or even thematic for them. But until that becomes the dominant *ethos*, we can expect that the remote achievements of Lonergan will continue to be chronically underestimated, too often dragooned into the service of academic production and academic advancement, and therefore more or less continually strained through the mesh of *haute vulgarization*. The question then becomes whether portions of the *haute vulgarization* can in some sense be redeemed. Lonergan once remarked that certain core discoveries on knowledge and reality in the history of classical and Christian thinking have to be liberated "from the endless incrustations and distortions contributed by simplifiers and vulgarisateurs from Isocrates and Cicero down to the still lingering remnants of classicist culture." 27410DTE070, 12. But the same rescue operation may also be necessary within Lonergan studies. See, e.g., *ibid.*, 20 ("It is not to be assumed that this invariant structure is as jejune as the triad: experience, understanding, judging.")

in *bavardage quotidian* or committing some variation on "the popular fallacy."[52] When the same man, writing within an almost identical context of unresolved problems stemming from a lack method in theology, and writing perhaps a month earlier, critiques a defective notion of theological understanding by asserting that it is "lacking in clarity and precision and so runs the risk of encouraging merely enthusiastic nebulosity,"[53] it is not likely that he himself is engaging in obliquity, imprecision, and enthusiastic nebulosity.

Lonergan, evidently, did not intend to run that risk in characterizing the perspective that for him was falling into place. He attempted instead to characterize his new perspective on the method of theology with considerable clarity and precision, if only to himself and a relatively narrow audience of contemporary or future readers. It could be, as well, that he was basically formulating his emerging perspective with clarity and precision primarily to himself and then only indirectly reporting that to Crowe in a kind of running account of his efforts-in-progress.[54] Yet however that may be, the meaning of his utterance is anything but vague, simple, obvious, or nebulous. It is a precise and novel meaning.

Still, "the greater the novelty, the less prepared the audience."[55] As I noted above, Fred Crowe was mystified by Lonergan's expression.[56] Was Lonergan's expression in the letter, then, a failure? "The expression will be a failure in the measure that insights B and D miscalculate the habitual intellectual development C

---

[52] *Insight, CWL* 3, 581.

[53] "Theology and Understanding," *Collection, CWL* 4, 123.

[54] This may be the most plausible scenario. The remarkable paragraph reproduced above is very oddly sandwiched between two interesting but otherwise unremarkable paragraphs. Without any warning, Lonergan suddenly shifts onto an entirely different level of expression. In the prior paragraph Lonergan reports on his reading of Mircea Eliade's *Image et Symboles* and discloses his own interest in putting together "a study of the significance of symbols as interpreting the content of the intellectual pattern of experience to the psyche (man as sensitive)," an interest that seems to have resulted in the subsequent insertion of a footnote in the already completed manuscript of *Insight. CWL* 3, 572, n. 7. It also mentions Odo Casel's *Mysteriengegenwart* and a commentary on it. The theme is mystery, but the tone is conversational. The following paragraph likewise reports in a conversational way on Lonergan's likely teaching duties the next year (*De Deo Trino*) and speculates on the texts he might use. Similar themes may be mentioned in the surrounding paragraphs—symbols as fostering internal communication with the sensitive psyche, intersubjectivity, mystery, Trinity—but nothing prepares the reader for the sudden escarpment (or perhaps it is an unexpected crevasse) between the paragraph on Eliade, with its conversational tone, and the immediately following paragraph reproduced above.

[55] *Insight, CWL* 3, 612. One has to wonder whether this effect contributed to greatly reducing the antecedent chances of a successful reception of the method of functional collaboration.

[56] See above, n. 41.

and the relevant deficiencies E of the anticipated reader."[57] An accurate answer to the question depends on how Lonergan conceived his anticipated readers' habitual intellectual development and relevant deficiencies in insight. Who was his anticipated audience for the utterance in question? The explicit and proximate audience was Fred Crowe, and we know that Fred did not grasp the deeper meaning and significance of the utterance. Did Lonergan "miscalculate"? It would seem that the expression was a failure.

On the other hand, perhaps Lonergan, as an "original genius," found "all current usage inept for his purposes"[58] and therefore felt the need to leap beyond current or customary usage. But in that case, by Lonergan's own account, his expression "succeeds remarkably if there is any possibility of grasping his meaning from his words."[59] This "any possibility" standard surely seems an oddly low bar with which to measure 'remarkable success.' Really, however, the standard measures the gap, sometimes an abyss, between a deeply original meaning or viewpoint and a contemporary receiving audience, between an accelerating meaning and the lagging cultural and linguistic resources available to express it and thereby communicate it. Only by deliberately torquing and straining current usage can an exceptionally original or creative thinker make current usage less inept a vehicle for expressing and communicating his or her original meaning. In the second sentence of the paragraph, Lonergan seems to have abandoned linguisticality and current usage entirely in favor of mathematically "available resources of expression."[60] At any rate, assuming the slim possibility Lonergan mentions, by this standard the expression would seem to be a success.

Successful or not, Lonergan's expression in the letter cannot be clear or precise to those lacking a mathematical background. So those lacking that background who aspire to be adequate readers of the letter have to acknowledge that we have much still to learn about Lonergan's conception of the method of theology. How much? A simple but telling crucial experiment suggests itself. When you read that phrase just now, "Lonergan's conception," did you find its meaning "simple and obvious"?[61] (Did I when I wrote it?) Did you more or less automatically and

---

[57] *Insight, CWL* 3, 579.

[58] *Verbum, CWL* 2, 37.

[59] *Verbum, CWL* 2, 37.

[60] *Insight, CWL* 3, 613. One should also think in this context of the complex mathematical analogy at the base of the key term, "operator," in Lonergan's account of development. *Insight, CWL* 3, 490. For its relation to the development of expression, see *Insight, CWL* 3, 597–98.

[61] One finds the meaning of expressions "obscure and difficult when he [the author or speaker] is stating what one has still to learn. In the latter case no amount of pedagogic and linguistic skill will eliminate the necessity of the effort to learn. For this reason only the

unconsciously sweep it into a neat and well-domesticated corner of standard learned academic discourse? Or did you notice and resist that "long-ingrained habit of thought and speech,"[62] deliberately displacing it into some wild known-unknown meaning, much like the equation-utterance in question?

For we cannot know his "conception," if we do not understand the equation; they are inconveniently correlative. In other words, what is lacking in us, as we struggle with the utterance, is likely to be not only the relevant technical mathematical background, together with what one might call "the heuristic turn" in conceiving conceptions in history,[63] but also any clear and precise "knowledge of all

---

man that understands everything already is in a position to demand that all meaning be simple and obvious to him." *Insight*, *CWL* 3, 581.

[62] *Method in Theology*, 239.

[63] "The heuristic turn" is important for many reasons, but one of them is its potential to transform old academic initiatives that through common acceptance have become settled, staid, stale, and self-perpetuating inertial routines into something resembling and supporting an orientation towards efficiently and functionally organized communities of inquirers. No science, no explanatory meaning, can emerge, survive, or flourish without precise heuristic structures. Not surprisingly, the heuristic turn is largely absent in the fields of ineffectual intellectual endeavor Lonergan calls "academic disciplines." *Method in Theology*, 3. The heuristic turn is not a shared notion, or even a shared slogan, among Lonergan students and scholars. On a larger scale, its absence haunts the long human present identified by McShane as the axial stage between compact consciousness and the future upper reaches of a developed interiority intimated by the symbol (about)[3]. For more on that symbol, see McShane, *Joistings* 1, 5–9, available at: http://www.philipmcshane.org/joistings. For its roots in Lonergan's technical expression in an early draft of chapter one of *Method*, see 47500DTE060, 8–9 (distinguishing first-, second-, and third-order consciousness and intentionality).

What would it be to conceive Lonergan's May 1954 conception of the method of theology in the full light of FS + UV + GS and the metawords, including W_3? See McShane, *The Allure of the Compelling Genius of History: Teaching Young Humans Humanity and Hope* (Vancouver: Axial Publishing, 2015), 95–101, and see also James Duffy, "Words, Diagrams, Heuristics," *Lonergan Gatherings* 7, available on www.academia.edu. Functional specialization, plus a seriously precise and heuristic view of the virtual potential totality of viewpoints, plus a developing genetic systematics, integrated into and under the guidance of a series of heuristic metawords, make possible a standard model in the human sciences; without a standard model scientific work simply does not occur.

The combination would endow those laboring in different functional specialties with the scientific benefit of a standard model, an integral heuristics. It would provide an increasingly refined and efficient grip on the geohistorical adventures and misadventures of human meaning in the stages of its historical unfolding. In the language of *Insight*'s way of envisioning this, it would provide a guiding "base in an adequate self-knowledge," an ongoing "retrospective expansion in the various genetic series of discoveries," and eventually "a dialectical expansion" cumulatively and progressively constituted by critically

that is lacking, and only gradually is that knowledge acquired."[64]

Let us then try to acquire an increment of the knowledge of what is lacking. An exercise in constructing a commonsense interpretation of exponential growth addressed to a typical commonsense audience might help disclose gaps in our knowledge and, more importantly, related gaps in our seeking or concrete capacities to seek. Imagine addressing an anticipated audience equipped with contemporary commonsense varieties of intellectual developments and deficiencies. In order to communicate to that audience, we might craft a flow of expression in light of what we understand to be the existing commonsense horizon in the audience to be addressed. The expression is designed, so to speak, to evoke an insight in the audience that is somehow loosely commensurate with both the thing to be understood and, as well, with the audience in its currently configured mesh of intellectual developments and deficiencies—or, as we might say, somehow commensurate with both the virtualities and the limitations of the anticipated audience's present horizon.

We would be in a position not unlike the position of the Grand Vizier Sissa ben Dahir addressing the Indian King Shirham. According to the legend, the king wanted to give Sissa a reward for inventing the game of chess. And so the king asked Sissa what reward he would like for so wonderful an invention. "Sissa addressed the king: 'Majesty, I would be happy if you were to give me a grain of

---

guided reformulations and reversals of "the many formulations of discoveries due to the polymorphic consciousness of man." *Insight, CWL* 3, 588–89.

This vital clue in *Insight* of Lonergan's heuristics of meaning in history finds confirmation in later passages. For example, "the genetic series of discoveries" and developments has to be itself cumulatively discovered, constructed, and revised. There is to be "the initial clearing of the ground" made possible by comparative method isolating "points of change to which all else reduces ... a clearing to be repeated as theological work advances, as its penetration of the historical process develops which draws attention to differences that may be developments." 86600DTE060, 7. The cumulative identification and seriation of developments, in turn, "means that your historical differences are put in an understood continuum, or you are trying to get them into an understood continuum or discontinuities, as the case may be. You are setting up relations, and by setting up the relations you move to understanding." *CWL* 17, 348. "Mappings and seriations" (2857ADTE070, 1) are developed and refined. Unlike present academic disciplines, the cycling does not tread aimlessly over the same old ground. Rather, the initial and repeated "clearing of the ground," together with functional cycling, will in time produce what Lonergan once called a "phylogenetic set of schemata" in different fields conducive to an ever more refined and comprehensive grip on "the stages and variations of human meanings, values, structures." Lonergan, "Philosophy and Theology," *A Second Collection*, ed. William Ryan and Bernard Tyrell (Philadelphia: The Westminster Press, 1974), 206. "The Standard Model means that there are 'diminishing returns.'" McShane, *Allure*, 182.
[64] *Insight, CWL* 3, 559.

wheat to place on the first square of the chessboard, and two grains of wheat to place on the second square, and four grains of what to place on the third, and eight grains of wheat to place on the fourth, and so on for the sixty-four squares.'"[65] The king was astonished at Sissa's apparent lack of ambition and intelligence. "And is that all you wish, Sissa, you fool?" But the relevant lack of intelligence resided more in the limitations of the king's commonsense horizon than in Sissa's requested reward. For the cumulative doubling of grains in each of 64 successive iterations yields something on the order of 18 quintillion grains, or about 200 cubic kilometers of wheat—enough to fill two billion railway cars.[66]

The progressive chessboard doubling is a geometric progression, involving a discrete sequence, while an exponential function is continuous. In other words, a geometric progression describes quantities having values at discrete points, while an exponential function involves quantities with values that change continuously. Nonetheless, the image of chessboard doubling suffices to evoke within commonsense horizons a noticeable tension between commonsense expectations and the reality of exponential growth. Geometric or exponential growth differs dramatically from ordinary and usual commonsense expectations of growth, whether those expectations might be derived from the growth of a plant, a tree, an animal, a food supply,[67] or some other source in the world accessible to ordinary common sense. Absent extended tutoring, our spontaneous expectations are not remotely geared to the reality to be understood.

So much for a commonsense apprehension of exponential growth addressed to a commonsense audience. The set of images and invited insights have a certain preparatory value. If you want, they assist the neuro-flexibility conducive to conceiving, within common sense, of fields beyond common sense. But Lonergan's expression in May 1954 of his perspective on the method of theology in terms of "evaluating $[1 + 1/n]^{nx}$ as $\underline{n}$ approaches infinity" is not an expression addressed to common sense from within the field of common sense. It is addressed and pitched beyond my present mathematical capacities, and so is beyond my present capacities to interpret.

Yet in a cyclic and cumulative enterprise, small, incremental, reliable steps matter, and it is possible for me to make a few restricted, largely negative judgments about its meaning. So, whatever else Lonergan's equation expression may mean, it does not stoop to *haute vulgarization*. It is not an instance of commonsense meaning.

---

[65] Clifford Pickover, *The Math Book* (New York: Sterling, 2009), 102.

[66] *Ibid.*

[67] See, for example, Thomas Malthus's insight into the different growth rates of populations and food supplies in his 1798 book, *An Essay on the Principle of Population, As It Affects the Future Improvement of Society.* Other things being equal, populations tend to increase by a geometric progression, while food supplies tend to increase only by an arithmetic progression.

It is not a vague or nebulous meaning, not a simple and obvious meaning, not a hasty or careless aside. It is on a different level of expression entirely, and it presupposes readers who recognize the need for, and the reality of, distinct levels of expression. It expresses a kind of capstone viewpoint on the road to a fuller solution to "basic and unsolved problems of theological method,"[68] problems that have bedeviled theology for centuries. It appears to be a precise, remote, complex, and evidently programmatic expression of a perspective Lonergan arrived at in the months following the completion of *Insight*.

No doubt these results of the interpretative effort seem disappointingly meager, thin gruel indeed if what one wants is a perspective parallel to the 49-year-old Lonergan's 1954 perspective on the method of theology. Yet as a would-be functional interpreter lacking the mathematical background necessary to apprehend the meaning and precision the expression has within the field of mathematics, I am not currently competent to give a more comprehensive or comprehending interpretation—far from it.

I hope that my effort is, in James Duffy's honest and challenging phrase, "a decent failure,"[69] but at the same time I have to personally join in "an honest effective acknowledgment in the present and next generation of Lonergan teachers that our background just did not prepare us for this shocking shift"[70] into explanatory meaning, including Lonergan's May 1954 perspective on the method of theology. My perspective on his perspective is simply inadequate. But isn't that, though, the point of the collaborative enterprise? "The collaboration of many contains a promise of success, where the unaided individual would have to despair."[71]

The key question is, where was (or is) this knowing going? Where does this crowning perspective of 1954 fit into a precise view of what was and is going forward in historical process, in the adventures and misadventures of human meaning and striving in history? Phil McShane's interlude attempts an answer—and with it, a more precise way of asking the question—from within the perspective of a functional historian. But first he completes the interpretation of the utterance, "and proceed to the limit as in evaluating $[1 + 1/n]^{nx}$ as $\underline{n}$ approaches infinity."

---

[68] *Collection, CWL* 4, 132.

[69] See James Duffy, "A Special Relation," infra, 92–93.

[70] Philip McShane, *Sane Economics and Fusionism* (Vancouver: Axial Publishing, 2010), 93.

[71] Bernard Lonergan, "The Original Preface of *Insight*," METHOD: *Journal of Lonergan Studies* 3:1 (1985), 3–7, at 6.

Interlude by Philip McShane

"Our spontaneous expectations are not remotely geared to the reality to be understood." The trouble, at this stage in our axial times, is the proximate gearing. That proximate gearing is a neurodynamic reality of the first world, caught in, trained into, an arrogance of initial meanings meshed richly together to disguise a stuntedness of culture that bows before technical competence in talk and technology. So, for instance, that proximate gearing could happily carry on with me talking here of, say, the daily growth of a young elephant as compared to the daily growth of an infant flea. Yes, there is a huge difference: the elephant adds the weight of pounds of adult fleas. Growth, yes, is a function of the starting point of the growing. See: are we not making progress towards understanding exponential growth, even enjoying looking at the equation, $d/dx\,(e^x) = e^x$? Even thinking of Lonergan's acceleration of growth between the end of *Insight* and the letter? And in that thinking miring ourselves in present deceits and conceits?

Our existential challenge, primarily ontic but also phyletic, in this stage of human history, is to begin to get a sufficient grip on our own meaning of our **what**, to generate in these coming centuries a culture that lives molecularly in the luminous regarding and guarding of the lack that is internal to infant humanity's cosmic puttering, recognized now, in this accepted Anthropocene Age, in its cosmic destructiveness. So, we are invited by Lonergan to gradually graduate, step up as a species, late in his book of invitations: "Most of all, what is lacking is a knowledge of all that is lacking and only gradually is that knowledge acquired."

Pat Brown's masterly presentation of the problem of interpreting Lonergan's short piece from 1954 nudges me towards brevity here: indeed, frankly, towards silence. Where does one begin when the bulk of present theological readers is not competent to read intelligently about the meaning of velocity, much less the velocity of the function e^x where e is the endless number 2.71828 ...? The oddness of the velocity of e^x is that it is a repeat of e^x, and this is quite obvious from the usual expression for e^x,[72] once one knows that $d/dx\,(x^n) = nx^{n-1}$. But what is that knowing? My experience of trying to help grade 11 and 12 schoolkids is that the present texts are dedicated to memorization as opposed to understanding, so the kids don't *know* that $d/dx\,(x^n) = nx^{n-1}$. Do you?

Here I note a curious wandering in my interlude. Pat obliquely points to the problem of the unpreparedness of present historians for Lonergan's briefly-expressed push in *Method*. That is the nudge behind these initial paragraphs of mine. So, immediately here, I turned to a pivotal component of understanding the text: the meaning of e^x and its velocity. But in doing that I jumped over problems of getting to e^x from the relevant piece of text, "evaluating $[1 + 1/n]^{nx}$ as n approaches

---

[72] $e^x = 1 + x + x^2/2 + x^3/2.3 + x^4/2.3.4 + \ldots + x^n/(2.3.4.5\ldots n) + \ldots$

infinity." Central to the jump is knowing, in a comprehending sense, that $[1 + 1/n]^{nx} = e^{nx\log(1 + 1/n)}$.

How far back, or forward, have I to go to bring you to a preliminary glimpse of this? Have you any memory of logarithms, indeed perhaps even a rote memory of "the logarithm of a number is the index of the power to which the base must be raised in order to equal the given number"? So, the 'log' of 1000 to the base 10 is? : well, $10^3 = 1000$. $\log_{10} 1000 = 3$. Got it? The little 10 simply names the base. In the text, the base, for Lonergan, is e. But how do we get to the equality mentioned? Might I push on with pedagogy? So, perhaps we could pause over the conditional, if $A = e^B$, then $\log_e A = B$. What might you say of the Log of A^5? Can you figure out—get help if not!—that $\log_e A^5 = 5B$? Next, notice, grasp, the simple jump to claiming—omitting the subscript e that designates the base—that $A^5 = e^{5B} = e^{5\log A}$. You have only to intelligently substitute $[1 + 1/n]$ for A, and nx for 5 and you grasp that you have arrived at the suggested equality. Are you genuinely, or vaguely in a genuine way, with me so far? Yet, what I have written so far here is all too dense to be decently pedagogical. So let me break off, break away, take off the brakes, take the baton as a functional historian, and locate Lonergan's "coming into perspective."

My perspective is a lean-to genetic one, a best present operable grip on "method in theology." It is a geohistorical heuristic of "Faith seeking understanding" that holds integrally *Genesis* and the *Upanishads*, the Zulus and the Zoroastrians and the Zen, and reaches hopily and hoppilly towards the eschatological seeking that is continuous yet startlingly discontinuous with the reach of "elementary intersubjectivity (cf. Sullivan's basic concept of interpersonal relations) to intersubjectivity in Christ," literally in the neurodynamic presence of all of us in the molecular delight that began with the billion-molecule mind of the infant Jesus.[73]

Into that perspective there arrives—let's pretend—freshly the paragraph of Lonergan sent to Crowe in 1954. My functional task is to luminously and effectively [74] fit this "coming into" into the present fullest geohistorical genetics. Not a task strange to Lonergan: there it is the minding of his typing, summer 1953, of what I call 60910.[75] It fits in too, with my perspective on his

---

[73] There is a vast and subtle literature on the prenatal and infant mind. Does it not have a place in the search for the historical Jesus? And are not His resurrected adult neuro-dynamics central to a serious eschatology?

[74] The first effectiveness is the successful exchange with Pat Brown here, but I would note that effectiveness is, in Lonergan's view, of the essence of both metaphysics and functional collaboration. The feeble statistics of actual implementation was a driving force in his eventual identification of Cosmopolis as a functional division of labour.

[75] 60910 refers to the paragraph at the turn of page 609 of *Insight*.

struggle, later that summer, with locating the treatise on the mystical body.[76] It fits in, further, within the perspective that blossomed out of his next ten and a half years of battling towards the differentiation of what he sketched in *Insight* earlier that same summer. There "are pure formulations if they proceed from an interpreter that grasps the universal viewpoint and if they are addressed to an audience that similarly grasps the universal viewpoint."[77] Further, the grip on effectiveness, a needy Faithfilled reality of Lonergan's psyche, luminous there certainly since the mid-1930s, has a blossoming fullness, within the full present perspective, of a street-reaching meshing of the promise of money with the Promise of the Covenant.[78] "For the rest: ordo universi," indeed.

My breakaway of those two paragraphs is like present summaries of Wiles' handling of Fermat's Last Theorem. Indeed, is there not a sweet parallel between Fermat's marginal scribble and Lonergan's paragraph? But best that I should break-back now to the struggle towards a precise interpretation of the text that fits in with the obsolete "academic disciplines" approach. We pause over the likely meaning Lonergan had for "in evaluating $[1 + 1/n]^{nx}$ as n approaches infinity." Had he in mind precise meanings of n and x when he wrote thus? Or was his view-bent simply a matter of illustrating a procedure to the limit, a growth in theology that was a generally accepted vague business? General vague business does not seem to square with his stand on "coming into perspective." Despite the year of teaching, or even because of it, he was in the high context of the unfinished climb talked of in the final chapter and epilogue of *Insight*. He seems in this 1954 text, to be coming to, pushing on to, a perspective, and indeed, giving the impression that he had arrived at some new height. Was it for him something of a definitive height? There is that impression from the affirmative bluntness of the text there, and from the manner in which he continued the text, down to the final scribbled comparison. Whatever his discovery, he places it unhesitatingly in the context of "ordo universi," and the scribbled second last word is a shout about the intelligibility of that ordo: "form."[79] Was he thinking of cosmopolis and of collaboration? Was he thinking of his problem regarding the

---

[76] His struggle is expressed in *Insight*, CWL 3, 763–64. My solution to his problem is to be found in *The Road to Religious Reality* (Vancouver: Axial Publishing, 2012), 18–22. It also solves the problem of the precise meaning of *Comparison*. See *Method in Theology*, 250.
[77] *Insight*, CWL 3, 602.
[78] A help here are the essays in *Divyadaan: Journal of Philosophy and Education*, vol. 21 (2010), "Do You Want a Sane Global Economy?," particularly my own two final essays there, "Edging towards a Later Global Economy," 233–244, and "The Global Economy and My Little Corner," 245–256.
[79] The shout is amusingly identifiable as something easy to hear. It is conveyed in a generic fashion in the final paragraph of *Insight* chapter 5, which begins with the phrase, "The answer is easily reached."

missing treatise on the mystical body? Was he thinking of his conclusion to his systematic treatise on the Trinity?

What we cannot easily doubt is that he was at some new high beyond these previous reaches. He was thinking of, or in, the *Imago* as it blossomed in his own mind and is to blossom in the theological community he was envisaging. That was his hope, within which there was an old conviction regarding the emergence of effectiveness.

What then might he have meant by *n*? Certainly it was a number. Was it a number relating to time or to people doing theology? Were these people, a core of all God's people, the "manifold of unities" he went on to write of? That would seem included. And what of *x*? For someone in his ballpark he was clearly talking about that curious function whose rate of growth was equal to itself: the function e^x, with $de^x/dx = e^x$. If so, there is the effort to communicate a view of theology, of the understanding of mysteries in this life—why not in the next?—as growing at the wondrous pace called *exponential*.

But that is not obvious if one is not in his ballpark. Is it obvious if one is in the ballpark presupposed by this volume as a slim heuristics? I do not think so: it would need a fattening of the heuristics bringing us into Lonergan's explanatory world. My *per se* functional interest here, however, coincident with Brown's effort, is some precise fattening of the heuristics of the emergent sub-group of functional historians, of the larger community to be identified as C_{ij}, and of the full community of common sense.

So my final comment here returns us to the two paragraphs of my outreaching **boldfaced** above. Let you reread them here, but now identified as within the present situation,[80] identified in the paragraphs beginning my interlude, identified as my backfiring to Brown, identified as positive *haute vulgarization*. How Now Brown Brow do you read these paragraphs and Lonergan's, how do you pick up and pass on an invitation to re-cycle Lonergan's little pointing?

---

[80] The word "situation" here can be taken to symbolize the massive difficulty of the transformation envisaged by Lonergan. It sits as a seven-fold nudge on page 358 of *Method in Theology*. *Allure* chapter 16 points to a lifting of that final chapter into a "coming into perspective" of a hierarchy of eight situation rooms with a complex topology that would ground a global effective "situation ethics." Efficient? That new complex structure represents "The Principle of Least Action" for future effective progress, and in using there the title of Feynman's 19th chapter of his *Lectures in Physics*, volume 2, I am bringing in a disturbing analogue of the difficulty of the project, a difficulty the prevalent commonsense perspective simply rejects. This is a massively difficult chapter in Feynman's apparently introductory work. You may recall a previous analogue from Husserl's early work in my chapter 3 on "The Calculus of Variation" in *Lack in the Beingstalk* (Halifax: Axial Publishing, 2006).

Thus I leave the topic, hand it back, backfire it, to Brown, as he moves to talk further of the problem of functional handing-on.

3. Handing Forward to Functional Historians

Earlier I mentioned Lonergan's problem of communicating within, and to, a deficient context, and connected it with what he calls "the variable standard of adequate expression."[81] The more profound and difficult the original meaning, the more acute and demanding becomes the problem of successful and adequate expression. "If one has anything very significant to say, then probably one will not be able to express the whole of it except to a rather specialized audience. Such limitations restrict the adequacy with which even one's principal meaning is expressed."[82]

Phil McShane's interlude has something very significant to say, a currently-remote principal meaning concerning the sophisticated heuristic perspective functional historians are to have in receiving the baton, the hand-off, from functional interpreters. He has something very significant to say, but the rather specialized audience equipped to hear the whole of it does not yet exist. His two-paragraph expression locating Lonergan's 'coming into perspective' is unsparing to readers more used to established literary modes of expression than to a meta-scientific level of expression.[83] Its remoteness therefore risks coming across as off-putting—not unlike Lonergan's paragraph.

But risks "coming across as off-putting" to whom? A sufficiently specialized audience is standard fare in any science. Specialized members of the mathematics community would be nonplussed by McShane's references to "The Calculus of Variation" and Wiles' proof of Fermat's Last Theorem. They might pause to bring their specialized knowledge to bear on his challenging paralleling of that theorem to Lonergan's paragraph. But they would hardly feel rankled, put-off, or put-upon. The same audience would find his discussion of "evaluating $[1 + 1/n]^{nx}$ as n approaches infinity" small beer or small ball, so to speak, in a very large ballpark. It is not that McShane's level of expression would be tolerated; it is that lesser levels would not be.

Where does that leave those of us who are not in the ballpark, those of us whose current capacities are limited, perhaps, to something closer to selling hotdogs outside? With the exercise of a little genuineness, it leaves us at least with a hard-earned, non-resenting respect for levels and achievements of expression with which

---

[81] *Insight*, *CWL* 3, 580; see above, n. 42.

[82] *Insight*, *CWL* 3, 580.

[83] *Insight*, *CWL* 3, 594 (noting that his classification of levels of expression leaves "abundant room for the introduction of further differentiations and nuances.")

we are not familiar. It should also leave us bereft of crude assumptions that in lesser mortals could easily sponsor a self-justifying round of *ressentiment*.

> To recognize the existence of levels of expression is to eliminate the crude assumptions of the interpreters, and still more of their critics, that take it for granted that all expression lies on a single level, namely, the psychological, literary, scientific, or philosophic level with which they happen to be most familiar.[84]

Maybe it leaves us also with some possibility of glimpsing the effective conditions of "the possibility of scientific collaboration, scientific control, and scientific advance towards commonly accepted results."[85] Failing that, it may leave us at the least with a glimmer of the pressing need for the kind of collaboration, control, and effective advance towards commonly accepted results characteristic of any empirical science but not of any present philosophy, theology, or other human science.

McShane asks me, tasks me, to answer two basic questions: "How do you read these paragraphs and Lonergan's, how do you pick up and pass on an invitation to re-cycle Lonergan's little pointing?" Here I am forced to pause long and hard on these two questions; more honestly, here I am forced to a grinding halt. How *do* I read those paragraphs, in my little home of wonder, way out of my league, in an unfamiliar ballpark, and me, a raggedy-assed mother-tucker from Seattle?[86] How *do* I conceive and envision "a rather specialized audience" of future functional historians to whom I would hand on an invitation to recycle "Lonergan's little pointing," and what form would the handing-on take? What *am* I to do with McShane's two-paragraph equivalent of summaries of Wiles' 108-page proof of Fermat's Last Theorem, a proof dauntingly incomprehensible to all but a highly specialized audience?[87]

---

[84] *Insight, CWL* 3, 594.

[85] *Insight, CWL* 3, 609.

[86] This seems like a very odd and sudden break with the usual standards of scholarly decorum. It is, in fact, merely an apt allusion to Bobby Tucker, a conductor, arranger, and friend of the composer Quincy Jones. One day, Tucker accompanied Jones to a house outside Paris for a master class on Stravinsky given by the 70-year-old Nadia Boulanger. His response to observing Boulanger conduct the class? It "just flipped me out. I'm just a little raggedy-ass motherfucker from Morristown, New Jersey, ain't I? What am I doing here?" McShane, *Allure*, 17 (quoting *Q: The Autobiography of Quincy Jones* (New York: Doubleday, 2001), 122–23). Not unlike my response to the two bolded paragraphs. And yours?

[87] I would encourage you to pause here and try the experiment of reading the first two or three paragraphs of Wiles' proof; like McShane's two paragraphs, they provide a bracing contrast between the literary and the scientific levels of expression. Andrew Wiles,

The first thing I need to do is re-read them, and I invite you to do the same, repeat footnotes and all.

My perspective is a lean-to genetic one, a best present operable grip on "method in theology." It is a geohistorical heuristic of "Faith seeking understanding" that holds integrally *Genesis* and the *Upanishads*, the Zulus and the Zoroastrians and the Zen, and reaches hopily and hoppilly towards the eschatological seeking that is continuous yet startlingly discontinuous with the reach of "elementary intersubjectivity (cf. Sullivan's basic concept of interpersonal relations) to intersubjectivity in Christ," literally in the neurodynamic presence of all of us in the molecular delight that began with the billion-molecule mind of the infant Jesus.[88]

Into that perspective there arrives—let's pretend—freshly the paragraph of Lonergan sent to Crowe in 1954. My functional task is to luminously and effectively[89] fit this "coming into" into the present fullest geohistorical genetics. Not a task strange to Lonergan: there it is the minding of his typing, summer 1953, of what I call 60910.[90] It fits in too, with my perspective on his struggle, later that summer, with locating the treatise on the mystical body.[91] It fits in, further, within the perspective that blossomed out of his next ten and a half years of battling towards the differentiation of what he sketched in *Insight* earlier that same summer. There "are pure formulations if they proceed from an interpreter that grasps the universal viewpoint and if they are addressed to an audience that similarly grasps the universal viewpoint."[92] Further, the grip on effectiveness, a needy Faithfilled reality of Lonergan's psyche, luminous there certainly since the mid-1930s, has a blossoming fullness, within the full present perspective, of a street-reaching meshing of

---

"Modular elliptic curves and Fermat's Last Theorem," *Annals of Mathematics*, 142 (1995), 443–551, available at: https://www.math.ias.edu/~anindya/fermat.pdf.

[88] There is a vast and subtle literature on the prenatal and infant mind. Does it not have a place in the search for the historical Jesus? And are not His resurrected adult neurodynamics central to a serious eschatology?

[89] The first effectiveness is the successful exchange with Pat Brown here, but I would note that effectiveness is, in Lonergan's view, of the essence of both metaphysics and functional collaboration. The feeble statistics of actual implementation was a driving force in his eventual identification of Cosmopolis as a functional division of labour.

[90] 60910 refers to the paragraph at the turn of page 609 of *Insight*.

[91] His struggle is expressed in *Insight*, CWL 3, 763–64. My solution to his problem is to be found in *The Road to Religious Reality* (Vancouver: Axial Publishing, 2012), 18–22. It also solves the problem of the precise meaning of *Comparison*. See *Method in Theology*, 250.

[92] CWL 3, 602.

the promise of money with the Promise of the Covenant.[93] "For the rest: ordo universi," indeed.

How do I read those paragraphs, and through them, Lonergan's paragraph? I am tempted to reply 'as, in the old joke, porcupines make love: very, very carefully.' More seriously, I read them, first, as texts on a scientific, not a literary or descriptive, level of expression. That is, I read them as the product of what I know to be the author's lifelong, relentless devotion to, and cherishing of, theoretical understanding.[94] In light of that devotion and cherishing, I read them with the special care and caution that comes from taking seriously the assertion that "all we know is somehow with us; it is present and operative within our knowing."[95] I read them, in short, "as one who has travelled and climbed and come down and who knows he has still harder going ahead 'reads' a map."[96] So, I attempt to read the paragraphs with the best *ethos* of climbing and map-reading I can muster.

Second, I read them as expressions of the heuristic turn and, in that sense, as an affirmation of the necessity for any scientific endeavor of complex heuristic structures, as well as a pointed and incarnate repudiation of the specious ideal of *voraussetzungslos* knowing, of viewpointless perspectives. The more complex the knowing, the more complex the heuristics necessary to evoke and express and expand it. Moreover, viewpoints develop ontogenetically and phylogenetically. A viewpoint on viewpoints will presuppose and integrate complex heuristics regarding both development and viewpoints.

Third, every insight and every resulting concept are at least potentially heuristic of further, future insights and expressions. But not all heuristics are created equal. The heuristic structures of empirical method function smoothly and efficiently on the basis of apt and efficient networks of symbolisms, schemata, diagrams. It doesn't take a rocket scientist to know that rocket science involves a different kind, a different flow, a different level of expression. And it's not limited to orbital or fluid mechanics. Page through any textbook in physics, chemistry, biology, or neuroscience, for example, and you will find flocks of equations and cascades of schematic symbolisms. The only alternative to schematic symbolisms of any complexity is an insanely inefficient kaleidoscope of circumlocutions, ever-shifting, always unwieldy, and never efficient. "A best present operable grip" on a method

---

[93] A help here are the essays in *Divyadaan: Journal of Philosophy and Education*, vol. 21 (2010), "Do You Want a Sane Global Economy?" particularly my own two final essays there, "Edging towards a Later Global Economy," 233–244, and "The Global Economy and My Little Corner," 245–256.

[94] *Insight, CWL* 3, 442.

[95] *Insight, CWL* 3, 303.

[96] Cid Corman, "Introduction," *Back Roads to Far Towns: Bashō's oku-no-hosomichi* (New York: The Ecco Press, 1996), 10.

takes for granted "the significance of symbolism"[97] presupposed by any modern science. Central terms like "perspective" and "geohistorical heuristic" are apt to have sigla-like density and compression of meaning and to function like key schematic symbolisms in the larger heuristic frameworks of an established science.

Fourth, then, I read the second and sixth words of the first sentence of McShane's first paragraph in the light of the heuristic turn in conceiving conceptions in history, and the fifth word of the second sentence of that paragraph in the same light. I read the fourth word of the second sentence, the adjective "geohistorical," in light of chapter 5 of *Insight*,[98] in light of the demand for a "retrospective expansion" of the universal viewpoint envisioned in *Insight*, which includes "the genetic sequence in which insights gradually are accumulated" by humankind,[99] and in light of the canon of complete explanation.[100] I read the same adjective in light of Lonergan's use of the phrase "mappings and seriations" in an early draft for his article on "Natural Right and Historical-mindedness,"[101] as well as his use of the same terms in the published version of "The Ongoing Genesis of Methods."[102] I read the same adjective in light of my largely initial-meaning-mediated grasp of Lonergan's account of emergent probability. I read the word "integrally" in the second sentence of the first paragraph in the light of my slowly developing understanding of the integral dynamic heuristic structure called "metaphysics" by Lonergan. (See also Lonergan's paragraph in the letter, lines 5–6.) I read these words in the light of my own self-reading, and so note that for me, these are largely merely initial meanings, compass headings from a map-in-the-making, whereas for McShane they are operable actual contexts.[103]

Fifth, the relevant heuristic structures are both precise and comprehensive; they cumulatively guide the discovery and development of the full "genetic series of

---

[97] See, e.g., *Insight, CWL* 3, 42; *ibid.*, 43 ("the symbolism constitutes a heuristic technique"); *ibid.*, 421 ("explicit metaphysics ... would consist in a symbolic indication of the total range of possible experience"); *ibid.*, 465.

[98] How, exactly? The answer is *not* easily reached. See *Insight, CWL* 3, 195 and n. 76 above.

[99] *Insight, CWL* 3, 609.

[100] *Insight, CWL* 3, 93.

[101] See n. 63 above for the citation to the archival material and a fuller context.

[102] Lonergan, "The Ongoing Genesis of Methods," *A Third Collection*, ed. Frederick Crowe (Mahwah, NJ: Paulist Press, 1985), 146–165, 150 (noting that particular or special methods "incorporate fresh stratagems, models, mappings, seriations").

[103] His actual context on a geohistorical perspective commenced at the latest in 1970. See the discussion of multiple Markov matrices in relation to Toynbee's *Study of History* in Franklin Fisher, "On the Analysis of History and the Interdependence of the Social Sciences," *Philosophy of Science*, 27:2 (1960), 147–158, cited in McShane, *Randomness, Statistics, and Emergence* (London: Macmillan & Co., 1970), 80, 237. See especially the quotation and discussion on 237.

discoveries"[104] and developments. One can read that phrase all too easily as though it were just ordinary prose, or one could, with difficulty, read it as a clue to the need for developing a genetic systematics. In any case, the heuristic called for is virtually comprehensive over the historical and geographical spread of human meanings, values, structures, "a geohistorical heuristic" that somehow "holds integrally" the whole human show. Its integration of past and present searchings, structurings, and orderings concerns humankind "as a concrete aggregate" of probably a hundred billion humans so far, "developing over time," some three or four million years, in fact, "where the locus of development and, so to speak, the synthetic bond is the emergence, expansion, differentiation, dialectic of meaning and of meaningful performance."[105]

To me that sentence by Lonergan, parceled into the quoted phrases, is alive with layers of latent heuristics, diagrams within diagrams honeycombing the homely words. Of course, it could be just another plodding sentence, if you're just another plodding reader. But the same man who discovered emergent probability as the immanent "design or order of this universe"[106] and of human history, also wrote that sentence.

Sixth, the design or order of this universe, "ordo universi," is emergent probability. As it is the order structuring the process of human history, functional historians ideally will have an operative grip on it, and not simply the slippery, ersatz grasp of a merely notional and nominal apprehension. McShane's way of "taking up the baton" as a functional historian rescues the word "history" from a crypto-literary level of expression and advances it to a meta-scientific level of expression fraught with all the complexity required to have an operable grip on what, in fact, is or was going forward.

What, after all, is a "real apprehension" of history? The young Lonergan thought it necessary to climb long and hard in order to arrive at an "analytic concept of history,"[107] itself only a heuristic device in the service of serious efforts at historical synthesis. Do you? For decades Lonergan took very seriously the question, What stands to general history as mathematics stands to physics; what will

---

[104] See n. 63 above.

[105] Lonergan, "The Transition from a Classicist World-view to Historical-mindedness," *A Second Collection*, 1–9, at 5–6. Note that Lonergan explicitly contrasts this apprehension of humankind with an abstract apprehension. *Ibid.*, 5.

[106] *Insight, CWL* 3, 139.

[107] See Lonergan, "Analytic Concept of History," *METHOD: Journal of Lonergan Studies* 11 (1993), 5–35; Patrick Brown, "System and History in Lonergan's Early Historical and Economic Manuscripts," *Journal of Macrodynamic Analysis* 1 (2001): 32–76, 40, n. 25; *ibid.*, 41–46; see also James Duffy, "Words, Diagrams, Heuristics," *Lonergan Gatherings* 7, available on www.academia.edu.

provide "history with heuristic structures,"[108] as mathematics provides heuristic structures for physics? Do we?

Finally, a caveat or caution. If McShane's two paragraphs really do parallel two-paragraph summaries of Wiles' proof, then there is no possibility here of my attempting anything other than the kind of brief and inadequate strategic pointers or intimations I have just attempted. Certainly I have very little to say about humanity's eschatological seekings and our destined intussusception into the molecular delight of Jesus.

There is no possibility of making a larger attempt, but also there is no need. McShane's paragraphs illustrate both a deeply needed precision on a perspective on perspectives in history and also the role of such a perspective in efficient, coordinated, and continuing collaboration. His paragraphs illustrate, objectify, reveal a conception and envisionment of the requirements for being a member of the "rather specialized audience" of future functional historians. It is not an easy envisionment to arrive at, and it is not an easy envisionment to communicate. It is not easily accessible to anyone, especially to those tutored in, and sheltered by, current versions of philosophy and theology. It is not easily domesticated. Nor is it easily assimilated to present rhythms of academic production and pressing concerns with academic advancement. But that, in part, is the point.

"My *per se* functional interest here, however, coincident with Brown's effort, is some precise fattening of the heuristics of the emergent sub-group of functional historians, of the larger community to be identified as C_{ij}, of the full community of common sense." Part of that "fattening" includes envisioning the functional specialties as a hierarchy of eight mutually interacting "situation rooms,"[109] looped and layered centers of reflection, implementation, and feedback floating over every concrete situation.

The extent to which one finds this an implausible over-reading of the thrust of Lonergan's method and project is, I believe, the extent to which one has radically

---

[108] *Method in Theology*, 141. The question occupied Lonergan intensively for decades. See 713-1EDTE030 (single typed page titled "Historical Analysis" from the 1930s)("Just as physical or chemical research presupposes a mathematics that largely is prior and independent, so too history presupposes the determination of the categories or pure correlations for which historical data can never do more than supply a content."); 713-1FDTE030 (single page of handwritten notes, also titled "Historical Analysis")("profound reading of literature, broad mastery of fact, keen perception of analogy, do not suffice.") See also *Topics in Education*, CWL 10, 251; "Natural Right and Historical-mindedness," *A Third Collection*, 178. It is fair, I think, to ask whether we are really up with the Lonergan of the late 1930s in the unwavering clarity of his unmistakable judgment concerning the forms of learning that "do not suffice," and in his ringing recognition of the need for complex heuristic structures. Or are we more or less tacitly content with the established routines of profound reading, broad mastery of fact, and keen analogies?
[109] See above, n. 79, and *Allure*, 66, n. 19; *ibid.*, 166, n. 19; *ibid.*, 192–97, and 193, n. 19.

under-read Lonergan. Either the method of functional collaboration was the culmination of his decades-long attempt to find an efficient form for informed, wise, concrete, resolute, and effective interventions in the dialectic of history, or it wasn't. But it seems that it was. For he himself speaks of the need for a "strategy" involving "a creative project emerging from a thorough understanding of a situation and a grasp of just what can be done about it."[110] The needed project is "an ongoing project constantly revised in the light of the feedback from its implementation."[111] It is not, in fact, a single project at all, "but a set of them, constantly reported to some central clearinghouse"[112] in order to harmonize the function of the different parts and to keep all parts informed of what has been achieved and "what has been tried and found wanting." The project, or set of projects, has all the earmarks of functional collaboration.

But his strategy and project were not well received. The greater the novelty, the less prepared the audience, and Lonergan's strategy and project were extremely novel. What, then, about the reception-history of the new method? It is a history of what hasn't happened, but should have. Was Lonergan's fear, voiced as part of his initial draftings of *Method in Theology* around 1966, something that was later realized in the historical flow? He thought it important to take the time and effort to project a scenario in which "the apprehension of method" is reduced to sloganeering, its "acceptance" is motivated by "the other-directedness of conventional minds," and, not yet adequately understood, "the rules of the game" serve merely to "safeguard the prestige and privileges of an in-group, but prevent rather than promote the advance of science."[113]

4. Concluding Reflections

My task has been to interpret, functionally, an obscure but important passage in a 1954 letter by Lonergan. The passage is obscure (a) because Lonergan compressed within its spare and unsparing manner and level of expression his complex and then-emerging perspective on the method of theology, and (b) because he used a level of mathematical analogy that is pitched over the heads of the vast majority of philosophers and theologians, then and now, including mine. It is obscure, not in itself, but only to us as meshed in our past and present limitations. One of the goals of successful functional interpretation and collaboration is to transcend those.

The passage is important for the same reasons that it seems obscure: it expresses a perspective on the method of theology by a highly original thinker—

---

[110] Lonergan, "The Response of the Jesuit as Priest and Apostle in the Modern World," *A Second Collection*, 187.
[111] *Ibid.*
[112] *Ibid.*
[113] 47500DTE060, 4.

truth be told, a thinker at or near the level of an Einstein or Aquinas, or rather, both—and it expresses that meaning in a highly difficult way, a way we are not used to, a way that is anything but familiar. He is, in effect, throwing down a gauntlet, and we are, in effect, expected to run the gauntlet. As he once noted, the audience he expects, the general ruck of anticipated audience (perhaps also "anticipated" in the sense of hoped-for), is an audience of people who "can grasp at the rudiments in calculus and analytic geometry, and so on, and are able to read with profit a book like Lindsay and Margenau's *Foundations of Physics*."[114] Take the time to riffle through a copy of Lindsay and Margenau and then ask yourself, Who did he have in mind? Who was he talking about?

He was talking about theologians and philosophers. And so we are confronted with the shock of a judgment that is contrary to the whole thrust of our present orientations, contrary to the present view we have of ourselves and of our roles and tasks, clean contrary also to the solution we may have embraced to the problem of living in a scholarly or academic environment, perhaps for decades. Does Lonergan's expectation "flip you out"? Are you pained or honest enough to ask, "What am I doing here?"[115] Or perhaps to ask, "What planet is he from?" Or even to ask, "What century were we just in?"[116] If this expectation makes you long for the good old days and ways of "initiatives that through common acceptance have become inertial routines,"[117] you are probably not alone. If it makes you long for a time when "profound reading of literature, broad mastery of fact, keen perception of analogy"[118] sufficed, you probably have a lot of company. But that would not be

---

[114] 89000DTE070, 3 (transcript of question and answer session, Boston College Lonergan Workshop, June 18, 1976); see also his notes for that session. 27920DTE070, 5.
[115] See above, n. 86.
[116] This was the question Lonergan asked Phil McShane upon leaving the Jesuit Provincial's residence in Dublin in 1961. See McShane, *Field Nocturnes Cantower* 117, page 7, n. 14. I am, of course, flipping it to the future here. Lonergan was in a different century, but unlike the Provincial, his wasn't in the past. See *ibid.*, section 1, "The Audience of the Universal Viewpoint," 3–6, http://www.philipmcshane.org/field-nocturnes-cantower.
[117] *Insight, CWL* 3, 559.
[118] 713-1FDTE030. The vast distance between this view of adequacy and Lonergan's own view is gauged in part by what we might call "the Lindsay and Margenau imperative," or perhaps better, "the Butterfield Drive." The latter is a marvelous symbolic message—see above, n. 54—that John Benton's neurons sent up in a dream several years ago. In the dream, he wanted to live at an address on Butterfield Drive, and when he awoke he had the obscure sense that the dream was about the challenge of taking seriously Herbert Butterfield's statement, endorsed frequently by Lonergan, that the scientific revolution "outshines everything since the rise of Christianity." Butterfield, *The Origins of Modern Science* (New York: The Free Press, 1957), 7. There is, indeed, "a drive to know." *Insight, CWL* 3, 28. And we are obligated not to suppress or impede it.

running the gauntlet or meeting the challenge. That would be fleeing the challenge and evading the gauntlet; that would be shirking the task; that would be placing yourself in league with the flight from understanding.

Like St. Paul, Lonergan expects you to run the good race. But he is not expecting you to run it alone. He is expecting you to run it in a relay, a circuit of structured and productive collaboration, a circuit briefly and compactly illustrated by the interlacing within this essay of my effort and Phil McShane's interlude, and more extensively by each contribution to this volume. "The collaboration of many contains a promise of success, where the unaided individual would have to despair."[119]

And it turns out Lonergan had that solution in mind all along, before he ever wrote the letter of May, 1954. As he had written some nine months earlier: "It is with the conditions, preliminary to an effective collaboration, that the present work is concerned."[120] The "present work" is *Insight*, and by extension or extrapolation, *Method in Theology*. But it is also the present work, the work of the present, the task at hand, the challenge of laboring to meet the conditions for effective collaboration, the structured and functional collaboration that cumulatively just might one day make human life livable, [121] just might lift "ordinary living in its concrete potentialities"[122] to new and unexpected heights.

But the vast distance also goes to the challenge of living on the level of the times, of sedulously avoiding in one's theology the near occasions of the sin of backwardness. Lonergan, "Dialectic of Authority," *A Third Collection*, ed. Frederick Crowe (New York: Paulist Press, 1985) 5–12, 8 (referring to the "sin of backwardness, of the cultures, the authorities, the individuals that fail to live on the level of their times.") As Lonergan put it in his notes for the June 18, 1976 afternoon session of the Boston College Lonergan Workshop, being able to "read with profit a book such as Lindsay and Margenau ... is very important if the theologian is to have a sound knowledge of ... the analogies that the sciences may provide for an understanding of mysteries." 27920DTE070, 5.

[119] Bernard Lonergan, "The Original Preface of *Insight*," METHOD: *Journal of Lonergan Studies* 3:1 (1985), 3–7, 6.

[120] *Ibid.*, 6. On the timing of the completion of the 'original preface,' see Frederick Crowe, "A Note on the Prefaces of *Insight*," METHOD: *Journal of Lonergan Studies* 3:1 (1985), 2.

[121] *Topics*, *CWL* 10, 232 (noting that philosophers for at least two centuries have been making an extraordinary mess of human life by promulgating "doctrines on politics, economics, education, and ... ever further doctrines" whose cumulative effect has been to drive us to the brink of unlivable lives).

[122] *Ibid.*, 208.

4. FUNCTIONAL HISTORY AND FUNCTIONAL HISTORIANS

Aaron Mundine

Part I. Cultivating a Relaxed Pace: SLOW DOWN![1]

I don't know all that I don't know about functional historians (who have yet to evolve), but one "thing" I know I know about them is that they will have a relaxed pace—something like, but much more developed than, the relaxed pace I've been trying to cultivate in myself these past four months since I headed out of the Boston College philosophy program a Mad Ape. It ain't easy trying to read and write to the beat of a different drummer since pretty much everyone is always going "a thousand feet per second"[2] (so, for example, I've "caught" myself walking too fast through the grocery store); and then there's getting and keeping a job so I can bring home some bacon—not to mention all the other needs of day-to-day living, like enjoying a little time with friends and family. But slowing down is worth the effort since slowing down always results in a lift not just of my reading and writing but of my living, too, and, what's more, bears the promise—with the collaboration of a series of contingents of slow-learners[3]—of lifting everyone else's living beyond "these

---

[1] It may be fun for you to know that for me SLOW symbolizes the first four functional specialties that "look back on" our slow development in time—and DOWN! symbolizes the final four functional specialties and C9, all of which "bring" bright ideas "down" to earth and town.

[2] That phrase comes from "The Tourist," the twelfth and final track on Radiohead's 1997 alternative-rock album *OK Computer*. While I'm at it, the subtitle for this part of the essay comes from the chorus of the track.

[3] The allusion here is to a collection of short stories written by Thomas Pynchon, a contemporary American author with a hearty appreciation of physics, chemistry, history, and education. In the introduction to the collection he writes, "Except for that succession of the criminally insane who have enjoyed power since 1945, including the power to do something about it, most of the rest of us poor sheep have always been stuck with simple, standard fear. I think we have all tried to deal with this slow escalation of our helplessness and terror in the few ways open to us, from not thinking about it to going crazy from it. Somewhere on this spectrum of impotence is writing fiction about it." *Slow Learner: Early Stories* (New York: Vintage, 2000), 18–19. This collection of stories offers, for me anyway, a complex, unconventional hope for a potent spectrum that in its very emergence will escalate us poor sheep beyond our present helplessness.

demon days"[4] to a few nice little places in the stars—after a few millennia of "cumulative and progressive" growin' up.[5]

How will functional historians—members of some future contingent of slow-learners whose evolutionary ancestors include me and, perhaps, you—contribute to that growin' up of history? Here—to date—is my guess about one of the numerous tasks of functional historians: Whether the view is of metaphysics or some methodology or some science or some art or some business or whatever apparently trifling problem you please, the work of functional historians will be to re-vise the genetic sequence of effective meanings[6] of that view in view of some functional interpreters' interpretation of the etch**ings** or **in**k-marks or pixel(em**en**t)s[7] or whatever encoding some funky-fresh or old-and-neglected "fruitful idea."[8]

---

[4] The phrase comes from the refrain of the title track, "Demon Days," off of the Gorillaz's 2005 album. I doubt that Damon Albarn and Jamie Hewlett thought this when they named the virtual alternative rock-rap-pop group, but I can't help but think of all of us alive these days as gorillaz in-between the two times of the human subject, "a" symbolizing the first time, and "z" symbolizing the second time. The two times are first mentioned on page 405 of *CWL* 12, *Triune God: Systematics*, but here I think the relevant quote comes from page 407: "in the state of fallen human nature, there are many obstacles that prevent temporal subjects from truly and genuinely becoming persons of the subsequent phase" [viz., the second time of the temporal subject].

[5] The allusions are to the second track, "Growin' Up," on Bruce Springsteen's first album *Greetings from Asbury Park, N.J.* (1973): those of us trying to get functional specialization rolling, among other "things," need the exuberance expressed in that track.

[6] What makes effective meaning effective? How about when meaning "hits the streets"?

[7] Why the bold facing of "in," "in," and "en"? In part, to nudge my extroverted readers out of the view that the print that is seen and read is "out there." Why might bolding "in," "in," and "en" nudge the reader out of the simplistic view that the print that is seen and read is "out there"? A clue comes from answering the question, What does bold facing do? Another clue comes from pages 184–85 of *CWL* 2, *Verbum: Word and Idea in Aquinas*. Further good clues come from chapter 5 of McShane's *Wealth of Self and Wealth of Nations*, available at: http://www.philipmcshane.org/published-books. A final clue, regarding my selection of "in," "in," and "en": this nudge is just as much a nudge out of the view that the print that is seen and read is "in here."

[8] The phrase alludes to two passages in *Insight*, "In each stage of the historical process, the facts are the social situation produced by the practical intelligence of the previous situation. Again, in each stage practical intelligence is engaged in grasping the concrete intelligibility and the immediate potentialities immanent in the facts. Finally, at each stage of the process, the general bias of common sense involves the disregard of timely and fruitful ideas; and this disregard not only excludes their implementation but also deprives subsequent stages both of the further ideas to which they would give rise and of the correction that they and their retinue would bring to the ideas that are implemented." *Insight, CWL* 3, 254. "[C]osmopolis is concerned to make operative the timely and fruitful ideas that otherwise are inoperative." *Ibid.*, 264.

Everyone in every functional specialty works with a genetic sequence "in mind," but it's the functional historians that revise those sequences.[9] Revise, but not invent—it seems to me. That leaves primitives—such as me and, again, perhaps you—the work of inventing such sequences so that—in the future—the odds of the evolution of functional historians might be more favorable and—in the distant future—history might get a lift from functional historians. Part of that inventing is learning a bit of science. Another part is understanding precisely both what one is doing when one is doing that bit of science and how that bit of science stands to one's operations in one's communities.[10] I have only begun to start that inventing. Still, I have ingested[11] a few bite-size bits of science, reflected somewhat on those chemical and physical processes, and understood vaguely both what I was doing in that ingesting and how those bits of science stand to my operations in my communities. So I'm getting into the habit of appreciating my nourishment, and one result of such appreciation is that I now appreciate page 20 of *Verbum* as offering the beginnings of a genetic sequence of metaphysics, which in turn serves as a beginning for all other genetic sequences.

So they've "worked" for me, but why are those the parts of inventing a genetic sequence? The following from Lonergan offers clues:

> The history of any particular discipline is in fact the history of its development. But this development, which would be the theme of a history, is not something simple and straightforward but something which occurs in a long series of various steps, errors, detours, and corrections. Now as one studies this movement he learns about this developmental process and so now possesses within himself an instance of that development which took place perhaps over several centuries. This can happen only if the person understands both his subject and the way he learned about it. Only then will he understand which elements in

---

[9] Functional historians revise per se. Concretely—per accidens—revisions can come from anywhere within the functional cycle. With this "in mind," consider footnotes 15 and 16.

[10] The allusions are to a short piece by Lonergan titled "The Genetic Circle," from a 1962 lecture. "That circle—the systematic exigence, the critical exigence, and the methodical exigence—is also a genetic process. One lives first of all in the world of community and then learns a bit of science and then reflects, is driven towards interiority to understand precisely what one is doing in science and how it stands to one's operations in the world of community. And that genetic process does not occur once. It occurs over and over again. One gets a certain grasp of science and is led onto certain points in the world of interiority. One finds that one has not got hold of everything, gets hold of something more, and so on. It is a process of spiraling upwards to an ever fuller view." *Early Works on Theological Method I, CWL* 22, 140.

[11] Does my use of the term "ingested" "throw you off"? If so, you really ought to, need to, read chapter 5 of McShane's *Wealth of Self and Wealth of Nations*.

the historical developmental process had to be understood before the others, which ones made for progress in understanding and which held it back, which elements really belong to the particular science and which do not, and which elements contain errors. Only then will he be able to tell at what point in the history of his subject there emerged new visions of the whole and when the first true system occurred, and when the transition took place from an earlier to a later systematic ordering, which systematization was simply an expansion of the former and which was radically new; what progressive transformation the whole subject underwent; how everything that was explained by the old systematization is now explained by the new one, along with many other things that the old one did not explain—the advances in physics, for example, by Einstein and Max Planck. Then and then alone will he be able to understand what factors favoured progress, what hindered it, and why, and so forth.

Clearly, therefore, the historian of any discipline has to have a thorough knowledge and understanding of the whole subject. And it is not enough that he understand it any way at all, but he must have a systematic understanding of it. For the precept, when applied to history, means that successive systems which have developed over a period of time have to be understood. The systematic understanding of a development ought to make use of an analogy with the development that takes places in the mind of the investigator who learns about the subject, and this interior development within the mind of the investigator ought to parallel the historical process by which the science itself developed.[12]

Einstein and Planck—lacking any obvious, mutant super-hero powers—aren't obviously evolutionary sports. Yet they are—and one might learn from them and, so, evolve, without adding any more to one's forty-six chromosomes. Still Einstein and Planck were far from functional history. Is it realistic for me to expect that I, with the proper nourishment, might go from inventing a genetic sequence to revising whatever primitive genetic sequencing I've invented-ingested? The answer is yes—but we're a long way from 9011 AD: so in a decade or two, I might have invented some rudimentary genetic sequence and, so, have evolved into a proto-historian—

---

[12] Quoted in Philip McShane, *The Road to Religious Reality* (Vancouver: Axial Publishing 2012), 36–37. As McShane notes, the quotation is from Michael G. Shield's 1990 translation of Lonergan's *"De Intellectu et Methodo"*: "Understanding and Method." A more recent translation by Michael Shields of the same passage appears in "Understanding and Method," *CWL* 23, *Early Works on Theological Method* 2, 175–77.

and a hundred years from now I might die a pretty good one,[13] having made some efforts at not only revising but understanding that revising of my genetic sequence— but the evolution of functional historians will still—even a hundred years from now—be a long time coming.

But I'm getting ahead of myself: it's a long, long, long time coming, but it won't ever arrive if we don't do anything to "tip the odds."[14] One way of "tipping the odds"—in addition to nourishing my own genetic sequence luminously—is pretending that I'm a functional historian talking to fellow functional historians about some fruitful functional interpretation of *whatever you please*.[15] In this case, what pleases me is the human mind, and the fruitful functional interpretation isn't a functional interpretation but McShane's *Posthumous* 3: "A Commentary on **Inside**." In pre-tending, I'll be offering an example of my guess as to how functional historians will contribute to the growin' up of history.

Part II. Taking Cranial Chemistry Seriously: Don't You Know You Got a Brain? (C_{33})

So far headway in the study of the human mind has been spotty since Aristotle. Aristotle rightly affirmed that the human mind understands the intelligibility in the imaginative or sensible presentation; and he correctly affirmed the link "between" our capacity to imagine and our capacities to sense and to retrosense as well as the link "between" those three capacities and our internal organs.

Since Aristotle—as I said, and you all know—headway has been spotty. There's Augustine of Hippo; there's John of Damascus; there's Thomas of Aquino—all of whom made headway with respect to what Aristotle affirmed so rightly about the human mind's act of understanding. After Scotus, however, the results of that headway were forgotten by most scientists and scholars. Come Descartes, the links correctly affirmed by Aristotle were called into question and, for some, the human mind became epiphenomenal. Thanks to the efforts of Bernard Lonergan, we've managed—to some extent—to recover those forgotten results and reaffirm Aristotle's affirmations.

---

[13] The point isn't so much about me as about the potential human lifespan: how long might the elders of the distant future live? Surely functional specialization will lead to healthier lifestyles.

[14] What does it mean to "tip the odds"? I don't have anything more than a commonsense understanding of its meaning, but I suspect that in a more comprehensive context than mine its meaning includes an understanding of pages 143–44 of *Insight*.

[15] Concretely, I am in the matrix of communications that McShane has "sketched" in various places: think C_{ij}. So, even though the per se strategy of cycling has me talking to fictional specialists in functional history (C_{33}) and functional dialectic (C_{34}), I'm at the same time trying to nudge you into "tipping the odds" of the evolution of all the functional species and encourage my fellow collaborators who are working hard for the distant future.

But today, thanks to the efforts of our friends in functional interpretation,[16] we have now been made aware of a way not only to extend that extent, as it were, but to surpass—at least, for the time being, in anticipation of understanding—the achievement that is Aristotle, and that way is taking the findings of the neuroscientists seriously and, so, incorporating them into our genetic sequence of the human mind.

Thus far in our sequencing we've got an analog to anatomy with our identifying and naming of what-questions, direct insights, is-questions, reflective insights, and so on and so forth; and we've got an analog to physiology with our relating the thirteen elements among themselves; but no further, in spite of the lingering questions about, say, that first element, the imaginative or sensible presentation.

The findings of the neuroscientists are not the way forward, but add some determinations to the next stage in our sequencing since—it seems to me and the functional interpreters—the next stage in the genetic sequencing of neuroscience "overlaps" with our next stage. Neuroscientists so far have achieved a descriptive differentiation of the different parts of the human brain; accumulated an understanding that relates the described parts to organic events, occurrences, and operations in the human brain; and have begun to transition from an understanding of the described parts as organs to an understanding of the conjugate forms that systematize the otherwise coincidental manifolds of chemical and physical processes by studying, among many "things," the main chemical regulators of intentional acts as well as the hydrodynamics of the cerebro-spinal fluid that—to mention just one of its functions—"rinses" the metabolic waste from our brains. No doubt, at this point, you all have page 489 of *Insight* "in mind."

As I've already mentioned, the "take" of our friends back in functional interpretation on these findings is that the next stage of our sequencing "overlaps" with the most recent stage of the neuroscientists' sequencing because, well, as you might've realized from my talk of analogs above, that genetic sequence on page 489ff of *Insight* applies just as much to our study of the human mind. Moreover, they add, neither neuroscientists nor functional historians with a "bent" towards intentionality analysis such as ourselves will make much headway till this "overlap" is acknowledged and we begin collaborating in understanding comprehensively the chemical and physical processes of our embrained minds.

I think the functional interpreters are right. I think our peculiar "bent" towards intentionality analysis has lead us to neglect the organic development that makes

---

[16] Each of us has such friends in mind. The friends I have "in mind" are not functional interpreters but proto-functional interpreters Bob Henman and Bill Zanardi. It's specimens like these that we ought to have "in mind" as we try to imagine and anticipate functional species. That said, there are plenty of other specimens in the larger zone of friends in the full scientific revolution and—larger still—that zone of friends that includes, for example, John of Damascus.

possible the psychic and intellectual development we spend so much time puzzling over without making much headway. Indeed, I think we will only make much headway once we've taken their "take" seriously.

(After much functional talk, we historians discovered that we all affirmed what was expressed in Part II above and, as usual, we "came up with" a statement of our stance on KOR[17] with respect to our latest genetic sequence—a statement to be "handed on" along with Part II and a suggestion as to the value of this latest genetic sequence to our friends over in functional dialectic. Again, I'm doing the talking.)

Part III. Handing on to Dialecticians: Trust Us, We Really Know What's Real (C₃₄)

Friends, the greatest concern of those of us over in functional history with respect to our latest revision of the genetic sequence of the study of the human mind is to "bring together" within "a larger whole" our viewpoint with the viewpoint of neuroscience without "bringing in" the extroversion of the neuroscientists, not to mention their neglect of GEM[18] and their muddling of common sense and theory, both of which "reveal" that the neuroscientists have yet to emerge from the axial period.

Of course it's up to y'all to judge not only whether we have done so successfully but whether we ourselves have overcome extroversion, "gotten the hang of" GEM, and "control over" our shifts from common sense to theory and back again. That said, we've attached our account of our latest revision of the genetic sequence of the study of the human mind—that, along with whatever copies of our previous work y'all have is perhaps enough material to "work with."

Now, as to the values of this latest genetic sequence of ours, we think its greatest value lies in the lift it might give to common sense and classrooms—all classrooms, not just those with courses on neuroscience or the human mind. Then there's the value that lies in the lift into functional specialization it might give to neuroscientists. Then there's the value that lies in the lift it might give to our own functional researchers whose focus is on the human mind. What other values there might be, we leave those for y'all to discern.

---

[17] What is knowing? What is objectivity? What is reality? *Insight*, *CWL* 3, 413.

[18] It seems to me apt to quote Lonergan's second definition of GEM: "Generalized empirical method operates on a combination of both the data of sense and the data of consciousness: it does not treat of objects without taking into account the corresponding subject; it does not treat of the subject's operations without taking into account the corresponding object." Lonergan, *A Third Collection*, ed. Frederick Crowe (New York: Paulist Press, 1985), 141. Why call it the second definition? See McShane's *Joistings* 21, "Research, Communications, Stages of Method," which offers a genetic sequence of views on GEM. This essay is available at: http://www.philipmcshane.org/joistings.

Part IV. Movin' Right Along: In Search of Good Times[19]

So pre-tending's one way of "tipping the odds" of the evolution of functional historians. But why does McShane's *Posthumous* 3: "A Commentary on **Inside**," and why might functional historians of *the human mind*, do the same? How does *Posthumous* 3 "tip the odds," and what's so special about functional historians of the human mind; why not, say, functional historians of human biology?

Well, it seems to me the odds are that most of my contemporary readers will have found their way here after having read something by Lonergan and, so, most likely will have a preoccupation with intentionality analysis but no anticipation of ever making real headway in understanding the chemical and physical processes from which their preoccupation emerges. The trouble is, no one will ever make real headway in understanding intentional acts without also making real headway in understanding the chemical and physical manifolds that such acts integrate.[20]

Perhaps you, now, however, are on the same page with me, page 489 of *Insight*? Or perhaps you are still stuck Meno-wise on page 6 of *Posthumous* 3? Those diagrams of the dynamics of knowing and doing, should they not be—for students of intentionality analysis—some "images of the relevant chemical and physical processes"? But might we do better than such diagrams? Lonergan thought so, and expressed it on the same page:

> ... there have to be invented appropriate symbolic images of the relevant chemical and physical processes; in these images there have to be grasped by insight the laws of the higher system that account for regularities beyond the range of physical and chemical explanation; from these laws there has to be constructed the flexible circle of schemes of recurrence in which the organism functions; finally, this flexible circle of schemes must be coincident with the related set of capacities-for-performance that previously was grasped in sensibly presented organs.[21]

Right now the flexible circle of schemes of recurrence in which each of us reads, writes, and lives hardly coincides with our related set of capacities-for-

---

[19] The allusion to the song sung by Kermit and Fozzie in *The Muppet Movie* (1979)—among other "things"—is an excuse to mention that academics need to overcome their uptightness about writing and embrace a more foot-loose and free-fancy approach: a lot more would get done in the academy if folks didn't spend so much time keeping to themselves and "polishing" their writings but, instead, accepted the fact that whatever they produce ought to be out-of-date pretty soon after presentation.

[20] No neuroscientists will ever make real headway in understanding the chemical and physical processes of the human brain without also making real headway in understanding the intentional acts "that account for regularities beyond the range of physical and chemical explanation." *Insight*, *CWL* 3, 489.

[21] *Insight*, *CWL* 3, 489

performance. How might we bring about a better coincidence? We can start by having more realistic expectations for ourselves, such as the expectations symbolized by FS + UV + GS and $H_3\{H_2[H_1(p_i;\ c_j;\ b_k;\ z_l;\ u_m;\ r_n)]\}$. We'd do better if, with such expectations, we set about—here and there, when we can, in our "spare time"[22]— luminously contributing to solutions to "apparently trifling problems" about "extremely simple things."[23] One not so simple thing is the image, such as the diagram of the dynamics of knowing. What are some "apparently trifling problems" about the image? I have to admit that all I've got right now is a compact anticipation expressed by the question: What chemical and physical processes make the image possible? So I've got a lot of inventing to do. I suspect you do too. What are we to do—how do we go from that compact anticipation to some "apparently trifling problems"? In the distant future, thanks to the functional historians, there will be "a systematic ordering that [began] with questions about the least complex kinds of variables and advanced toward questions about more complex kinds"[24]—so the "hungry freaks"[25] of the distant future won't be malnourished like us. But what about us, now, in the distant past? McShane's website comes to mind, where one might find "trifling problems" about the physics of cups in motion and the chemistry of stomach acid. Such trifling problems might be a start for you towards a few trifling problems about the image's chemical and physical processes. However, they could just as well be a start towards a few "trifling problems" about anything and everything, since anything and everything meshes physically and chemically— including intentionality. You, then, unless you've only a casual interest in academic labor, have no honest way around such starts.

---

[22] The allusion is to a letter written by an over-worked, thirty-year-old Lonergan to a Jesuit superior: "What on earth is to be done? I have done all that can be done in spare time and without special opportunities to have contact with those capable of guiding and directing me as well as to read the oceans of books that I would have to read were I to publish stuff that is really worth-while." Bernard Lonergan, Letter to his Jesuit Provincial dated January 22, 1935, published in Pierrot Lambert and Philip McShane, *Bernard Lonergan: His Life and Leading Ideas* (Vancouver: Axial Publishing, 2010), 154.

[23] The *Insight* passages to which the phrases allude are worth quoting: "In the midst of that vast and profound stirring of human minds which we name the Renaissance, Descartes was convinced that too many people felt it beneath them to direct their effort to apparently trifling problems." *Insight*, CWL 3, 27; see also *ibid.*, 60 ("With another bow, then, to Descartes's insistence on understanding extremely simple things ...")

[24] William Zanardi, *Comparative Interpretation: A Primer* (Austin: Forty Acres Press, 2013), 272.

[25] The phrase comes from the opening track on the strange 1966 album by The Mothers of Invention, *Freak Out*: "Mister America/ Walk on by/ Your schools that do not teach/ Mister America/ Walk on by/ The minds that won't be reached/ Mister America/ Try to hide/ The emptiness that's you inside/ When once you find that the way you lied/ And all the corny tricks you tried/ Will not forestall the rising tide of/ Hungry freaks, daddy!"

I have been endeavoring to persuade you that meaning is an important part of human living. I wish now to add that reflection on meaning and the consequent control of meaning are still more important.[26]

For [humans] can discover emergent probability; [we] can work out the manner in which prior insights and decisions determine the possibilities and probabilities of later insights and decisions; [we] can guide [our] present decisions in the light of their influence on future insights and decisions; finally, this control of the emergent probability of the future can be exercised not only by the individual choosing his [or her] career and in forming his [or her] character, not only by adults in educating the younger generation, but also by [humanity] in its consciousness of its responsibility to the future of [humanity]. Just as technical, economic, and political development gives [humanity] a dominion over nature, so also the advance of knowledge creates and demands a human contribution to the control of human history.[27]

But what is the immediately pragmatic point? It is that we must tune forward towards that element of the second time of the human subject as best we can, in proleptic reachings. Part of that reaching is a detection of methodological reorientations and distinctions. Libraries and conventions and parochialisms of interest and concern cut us off from envisaging much less attempting collaborative and efficient minding. Nazi Germany is an organism of the past, but American, Chinese, and Catholic imperialisms are alive and narrowly well in prolonging the longer cycle of decline. Are we to rely on accidental reshufflings?[28]

---

[26] Lonergan, "Dimensions of Meaning," in *CWL* 4, *Collection*, 235.

[27] *Insight*, *CWL* 3, 252–53.

[28] Philip McShane, *Method in Theology: Revisions and Implementations* (2007), 47, available at: http://www.philipmcshane.org/method-in-theology-revisions-and-implementations.

James Duffy

I cherish the day
You're ruling the way that I move
You take my air
You show me how deep love can be[1]

I. Contexts

The context of my essay is a "somehow with me"[2] book or series of books to be written, the sketchy outline of which can be found in the Appendix to this essay. Here I will limit myself to describing three overlapping contexts

A. "Mis Queridos"

Some five years ago I decided to write a series of letters for students in a bachelor's degree in humanities and social sciences. The letters, written in Spanish, begin "Mis queridos," which translates "My dear ones," "My loved ones," or "My beloveds." Unlike most letters, these are titled and have footnotes. They are an unorthodox communication with students, with titles that intimate a certain eclecticism: "Grow, Baby, Grow"; "All You Need is Love?"; "The Salsa and Progress"; "What is Money?"; "Before Money: Acceleration and Self-Love";[3] and "El Chingón."[4]

What was I trying to do? Invite, cajole, and caress students, hoping for a union leading to a believing leading a few, perhaps, to one day understanding.[5]

The letters are a methodological mess crying out for differentiations of later days when the problem of truth[6] is vibrantly and luminously lived in classrooms, when current seeding of seedling efforts begin to bloom and bear fruit in mining

---

[1] Sade, "Cherish the Day."
[2] "All we know is somehow with us; it is present and operative within our knowing; but it lurks behind the scenes, and it reveals itself only in the exactitude with which each minor increment to our knowing is effected." *Insight, CWL* 3, 303.
[3] An adaptation of Terrance Quinn, "The Calculus Campaign," *Journal of Macrodynamic Analysis*, 2, 2002, http://mun.ca/jmda
[4] It is not easy to translate this phrase. "El chingón" is the person who "chingar"—to fail, to screw, to joke, to drink too much, to fuck (with), etc. This letter is, in part, a commentary on Octavio Paz, *El Laberinto de la Soledad*, chapter 4, "Los Hijos de la Malinche."
[5] See *CWL* 12, *The Triune God: Systematics*, 407.
[6] See "The Problem," *Insight, CWL* 3, 585–587. Notice how Lonergan shifts the problem of interpreting a text to the problem of teaching. How well has teacher James or Janette appropriated the three-step procedure, "A to F"? Read it and weep. In the third step the audience has a decent notion of its own intellectual development! Read it and laugh.

local mansions of meaning[7] and out-tonguing in words better than tongues to "help the whole church."[8] In addition, they are personal data on the rare "growth curve" $d/dx(e^x) = e^x$,[9] as well as intimations of missing levels of mediation that functional collaboration is going to provide when Maria de Los Angelitos[10] grows in wisdom, age, and grace. But, I believe, the letters were an honest effort to overcome stage fright—the fright of direct speech, which is usually more of a problem for academicians with graduate degrees than for undergraduates.[11]

B. A Decent Failure

From January 2011 until February 2012 a handful of brave souls attempted the functional specialties in an on-line seminar.[12] Because of the usual demands on our unlivable lives, there was neither the time nor the energy to finish the first of three

---

[7] See the contribution of Philip McShane "Foundations of Communications," at pages 163, 167, 171.

[8] "The man who speaks in strange tongues helps only himself, but the one who speaks God's message helps the whole church." 1 *Corinthians*, 14:4. See also section 4 of the last chapter in *Method*: "The Christian Church and its Contemporary Situation."

[9] $d/dx(e^x) = e^x$ $e^x = 1 + \frac{x}{1!} + \frac{x^2}{2!} + \frac{x^3}{3!} + \cdots$, $-\infty < x < \infty$.

[10] In three of these letters I reach in fantasy toward a local wise dame—her name is "Maria de los Angelitos." She is "una mujer culta" (an educated woman, "sufficiently cultured" [*Insight*, *CWL* 3, 22]) who benefits from having the best genetic systematics passed on to her, and who is just a mile or two—or a phone call—away from the university where I teach. Students and even a few colleagues now jokingly ask me how Maria is doing. The Spanish essay is available at: http://eltoquehumano-humanistas.blogspot.mx/2012/09/maria-de-los-angelitos.html.

[11] There is no going back now: some undergraduates, even a few graduate students, have confided that they prefer reading my mad rambles to Thomas Kuhn, Edgar Morin, Bernard Lonergan, Jürgen Habermas, Jean Piaget et al. Conventional name-dropping in comparative work, so much a part of current academic culture, will eventually be replaced by scientific elimination of names (exceptions would be mentioning such topics as "Mendeleev's periodic table"), in a special way in *Comparison*. Aristotle, Aquinas, Hume, and Hegel are in different time-tubes, with the former guys not benefitting from a reading of Hegel. Normally we do not compare their characters, although that too is relevant (see "Understanding the Author," *Method in Theology*, 160-161); normally the topic is something "real," some object "x." The one comparing should be in control of what he or she means by "real" and the best opinion about "x." The question, "Did Hegel make any progress on Hume?" implies a horizon beyond both of them, and indeed a position on progress, which is a hairy topic if the good is a history. See also note 15 below.

[12] Essays are available on-line at: http://www.sgeme.org/BlogEngine/archive.aspx.

cycles.[13] We ran out of gas in the fifth seminar, which was an effort at doing Foundations by way of comparison.

My own effort to spontaneously compare two houses ended in a quite humbling realization that such a comparison would require a wholesome heuristics of house and indeed a heuristics of "what?" Comparing spontaneously— improvisation, no looking back or sideways, addressing the audience as would a jazz musician[14] or stand-up comedian who is on a roll—sounded easier than it is. But the effort was not in vain; it intimated both the weirdness of Foundations as a no-looking-back fantasy and the difficult identification of the scientific meaning of *Comparison* as opposed to the usual merely academic meaning.[15]

C. Subjects in love

A third context is trifold: (i) the expression "subject as subject," which is a turn of phrase in *Phenomenology and Logic*[16] that is central for understanding the analogy between temporal and eternal subjects;[17] (ii) the phrase "being in love with God," a phrase which recurs in *Method in Theology*;[18] (iii) McShane's writings that weave around a single paragraph from *The Triune God: Systematics*.[19]

---

[13] The seminar was going to run from 2011 until 2017, in three cycles: (i) seminars 1–8 on the functional collaboration in each of the specialties with a focus on the general categories; (ii) seminars 9–16 on functional collaboration in the special categories of Christian thinking, (iii) seminars 17–24 on collaboration in a global context. A final seminar 25 Seminar would face the task of an integral eschatological perspective.

[14] Miles Davis, John Coltrane and Charlie Parker do not mysteriously create everything or create from nothing in the moment, but play something familiar for the first time and in so doing reorganize jazz molecules. "Improvisation is sped up composition and putting together things you already know in an artistic way." John Scofield.

[15] "That identification is to serve, in these coming decades, as a measure of the weakness and folly of present conventions that tie the meaning of comparison to an old-style comparing of X and Y, whether X and Y be giant periods of history or small individual strugglers towards meaning." Philip McShane, *FuSe* 22, "The 2012 Crisis of Speaking to the Future." See also Q. 38 "Comparing Jones and Smith," Q. 39 "Comparing Lonergan and Jones," Q. 40 "Self-luminous Comparing Work," and Q. 41 "Promoting Cyclic Comparing Work". This Q&A series is available at: http://www.philipmcshane.org/questions-and-answers.

[16] *CWL* 18, *Phenomenology and Logic*, 226, 314–17, 360–65.

[17] "The analogy, then, about which we are inquiring is the analogy of the subject as subject; for a temporal subject as well as an eternal subject is a distinct subsistent in an intellectual nature, but a temporal subject and an eternal subject are related to their respective intellectual natures in different ways." *CWL* 12, 401.

[18] *Method in Theology*, 39, 84, 105-109, 113, 242.

[19] The paragraph, cited below in section 2, begins on the bottom of *The Triune God: Systematics*, *CWL* 12, page 471 and continues on page 473. McShane's efforts to come to grips with the dense meaning of this paragraph are found in chapter 7, "Grace: The Final

Here I note an overlap of the three contexts, for the challenge of addressing as subject students as subjects is implicit in my letters, in the fifth e-Seminar on Foundations, and in faithfully seeking to understand the subject James in love with three Subjects.

Who is, who are, my audience in this contribution to *Seeding Global Collaboration*? It is a mixed group, and like my fellow authors in this volume, I pretend to precisely address. In my case I pretend to be capable of addressing: (i) those struggling to express doctrines; (ii) foundational characters, including myself as foundational character; (iii) the entire group of collaborators. I am a beginner, pretending to step onto the stage and step forward, in some little way, from the pointers given by Lonergan and McShane regarding a cultural shift towards living luminously in the subjectivity of God, a "stage of meaning when the world of interiority has been made the explicit ground of the worlds of theory and of commons sense."[20] The hubris of the past fifty years of Lonerganism would have me believe that citing a text from *CWL* which names a stage of meaning could somehow substitute for the very slow and dreaded conversion and ongoing repentance to a radically new, beautiful, and efficient embracing of students and neighbors.[21]

II. Content: The Meaning of *Cherishing*

The five words "Clasping," "Cherishing," "Calling," "Craving" and "Christing"[22] emerged for McShane after focusing for some three and a half years on a single paragraph in *The Triune God: Systematics*:

> But if one asks about the supernatural character of the formal terms, it is pertinent to note the following. First, there are four real divine relations, really identical with the divine substance, and therefore there are four

---

Frontier," *The Redress of Poise, ChrISt in History*, and the *Posthumous* series. Also relevant are the *FuSe* essays from the e-Seminar which ended where I begin in this essay. *FuSe* 22, "The 2012 Crisis of Speaking to the Future," is particularly relevant. The *FuSe* series is available at: http://www.philipmcshane.org/fuse.

[20] *Method in Theology*, 107.

[21] 1 *John* 4:20 invites axial repentance: "Whoever claims to love God yet hates a brother or sister is a liar." It was oh so easy to reply to my niece's little-flower question "Why are you crying?" See QQ. 21, 22, 23, and 24, "The Chemistry of the Searcher, Searching with a Nine-year Old, Craving Effective Transpositions, Who are You Three?" available at: http://www.philipmcshane.org/questions-and-answers.

[22] See *Posthumous* 11, 12, 13, 14, 16, 19, and 21. This *Posthumous* series is available at: http://www.philipmcshane.org/posthumous. The central essay in the series is *Posthumous* 11 "Allurexperiences," where McShane focuses on the textual sources of "religious experience" and "struggles to bring forward elements of Lonergan's view of religion and of the special categories needed to foster the global climb towards the related identity." *Posthumous* 11, 1.

very special modes that ground the external imitation of the divine substance. Next, there are four absolutely supernatural realities, which are never found uninformed, namely, the secondary act of existence of the incarnation, sanctifying grace, the habit of charity, and the light of glory. It would not be inappropriate, therefore, to say that the secondary act of existence of the incarnation is a created participation of paternity, and so has a special relation to the Son; that sanctifying grace is a participation of active spiration, and so has a special relation to the Holy Spirit; that the habit of charity is a participation of passive spiration, and so has a special relation to the Father and the Son; and that the light of glory is a participation of sonship, and so in a most perfect way brings the children of adoption back to the Father.[23]

The quote is extremely dense, for it is about primary relativities and secondary determinations. Four real divine relations are primary relativities, "four very special modes that ground the external imitation"[24] and participation in eternal mind in accord with schedules of probabilities. Here, now, I type along in a secondary determination that is "constitutive neither of the relation nor of its reality as relation."[25] There, then, you read along as secondarily contingent determinations of a nonsystematic manifold. And, *aber natürlich, desde luego*, of course my horse, we are in the dark.[26]

---

[23] The *Triune God: Systematics, CWL* 12, 471, and 473. (hereafter 47173). In a correspondence of April 24, 2014, McShane expressed a metadoctrine in these words: "Most simply, the existential Trinitarian reality is the human person meshed in 'accidental' Accepting of 'substantial' Calling. The tension between A and C 'processes' the person's story within the Story of Cherishing. A 4-person team-effort unique to this human person yet galactically focused within the being of all galaxies. The closeness of the team, three infinite P + micro-p, depends on the relaxed openness of the micro-p. The openness is to the being of all galaxies within circumincessional Being, an openness to 100+ billion of other micro-ps. Thus, theoretical understanding seeks to embrace, by letting Embracing loose in glocal molecules."

[24] "What is named internal relation in *Divinarum Personarum conceptio analogica*, p 272 ff. [*De Deo Trino: Pars systematica*, 291–315; see appendix 3, 687–737] and primary relativity in *Insight* [1957], p. 491 ff. [*Insight, CWL* 3: 514–520]." Letter to Fr Gerard Smith, S.J.," *The Triune God: Systematics, CWL* 12, 739.

[25] *Insight, CWL* 3, 519.

[26] If the real is already-out-there or already-in-here, then real divine and human relations are a fancy of the imagination, and not real at all. Likewise, if the "real" is seen with spiritual eyes, then looking at the relativity of a relation and looking at the contingent that is subject to manifold variations are irreducible. With one look I see or imagine a primary relativity; with a second look I see or imagine the secondary determinations that modify the absolute. Then I jump to the conclusion that the reality of the changing base and the reality of a change in the relativity must be two really distinct entities. On the luminously dark

I will focus on a single statement in this paragraph: "the secondary act of existence of the incarnation is a created participation of paternity and so has a special relation to the Son." What do these twenty-three words mean? What really happened? What is the historical lift? How does an infinite gift to finitude set fire to history? How does the "match" work: infinite gift matched, factually isomorphic with a contingent truth? Is it possible that, in a sense, the four real relations and their participations "say it all"?[27]

It would be a blunder to claim that the secondary act of existence is an isolated, act-tag-on to Jesus, a wriggle in the womb of a young Jewess. A new flower in the garden changes the garden; a new cry from a creature deep in the ocean changes the sea. Indeed does not a new flower in the garden or a new cry of a sea creature also change the cosmos? The odd reality that is the contingent term, the Jesus act, the primary *Cherishing*, is an "in-plant" in history that somehow changes the cosmos, and gives voice to the silent Word.

Cherishing names a grace of creative union, of everything becoming one, a meshing for us, through us, and in us. Teilhard de Chardin writes of this becoming one in *Science and Christ*:

> And in this rich and living ambience, the attributes, seemingly the most contradictory, of attachment and detachment, of action and contemplation, of the one and the multiple, of spirit and matter, are reconciled without difficulty in conformity with the designs of creative union: everything becomes one by becoming itself.[28]

Everything becoming one by becoming itself sounds a bit odd, a bit off, a bit too loose, unless Chardin had in mind intentional existence. On the other hand, everyone becoming one by actualizing *potens omnia facere et fieri*, and in a particular and

---

position of identifying is-ing what as performance, supposing that there are terms without relations or relations without terms, is (self) empirically identified, performed, and possessed as a blunder. "Identification is performance. Its effect is to make one possess the insight as one's own, to be assured of one's use of it, to be familiar with the range of its relevance." *Insight, CWL* 3, 582.

[27] A context is McShane's taking up Robert Doran's question: "Is the four-point hypothesis adequate on its own as a unified field structure for systematics?" in *Method in Theology: Revisions and Implementations,* "Part Three: Structure and Anticipations," available at: http://www.philipmcshane.org/method-in-theology-revisions-and-implementations, McShane suggests considering "the shocking integrative achievement of Maxwell's four equations" as something of an analogue. "Is there not a deep clue here regarding a genetic axiomatics that holds heuristically and heartily to the relations talked about as thus infinitely comprehensive of this history and all possible histories?" *Ibid.,* 175.

[28] Teilhard de Chardin, *Science and Christ,* translated from the French by Rene Hague (London: Collins, 1965), 74.

strange way by being factually, cosmically uplifted, might not sound so odd if one understands as Lonergan understood.

Questions about the one and the many, notional, real, problematic and mixed distinctions and relations, and the problem of external and internal relations are taken up by Lonergan in chapter 16 of *Insight*. He writes of "the relation 'father'" in the third line of the second section, "Relations." The chapter's title is "Metaphysics as Science." What, we might ask—and it is an asking that is the heart of my contribution—is the meaning for me or you of "as" in this title? Does its meaning merge with the meaning of "as" in "subject as subject"? Does its meaning merge with the "as" in "metaphysics as possible-to-me"? Would it be entirely outlandish to ask, then, about the meaning of "as" in "Metaphysics as Science of subjects as subjects in love with Subjects as Subjects"? Perhaps, "to meet this problem it is necessary to distinguish in concrete relations between two components, namely, a primary relativity, and other, secondary determinations."[29]

Here is where I come to the core of my advance on the meaning of *Cherishing*, McShane's name for the base, a cosmic base of my typing and your reading, a cosmic base that has a date. "The secondary act of existence of the incarnation is a created participation of paternity"—"an alliance and a love that, so to speak, bring God too close to man."[30]

"Draw me unto you and let us run together." (*Song of Songs* 1:4)

Heaven hounds little me and little you, making us leap far beyond the planets visited by the Little Prince. Three Persons have all secondarily determined stumbling, fumbling collaborators in eternal mind as we plod our way towards seeding a science of the Rose of Sharon in these pre-adolescent times.

This pivotal reality of finitude caresses my slow growth as a performer who craves an image of adult performances. I write from a view and from a craving of adult growth that, in good time, will give a secondary determination to the primary relativity that is described in chapter 5 of *Method in Theology*—homing "birds of a feather" at home in Maxwell's equations developing exponentially and in relation, "a manifold of developing and linked unities,"[31] "a new and higher collaboration of intellects through faith,"[32] street and school performers, then, who live and breathe

---

[29] *Ibid.*, 515.

[30] *Insight*, *CWL* 3, 747 (hereafter 747).

[31] See lines 4-5 of the paragraph in Lonergan's letter to Crowe from May, 1954, reproduced above at page 49.

[32] See note 8 of "Foundational Prayer II: All Saints' Reaching," *Prehumous* 5 (http://www.philipmcshane.org/prehumous) and *Insight*, 745.

and move in functional global care as scientists of He who seekest fondest, blindest, weakest little me.[33]

The Divine cosmic lift towards the absolutely supernatural is a three-in-one act that "involves a true and full identity between the love and beloved," both a psychological and ontological indwelling (circumincession).[34] Three Persons have "my universe"[35] in eternal mind, while I, in "collaboration on the part of many,"[36] faithfully-thinkingly not only imitate Their order "but also in some manner participate in that order."[37]

III. Hand-on

Handing on is an element of my would-be Foundational efforts in a special way. McShane talks regularly of the two functions of Foundations: fantasy and acceleration.[38] Here I will comment upon two distinct hand-ons: (i) the per se C_{56} conversation; (ii) the C_{5x} conversation.

"Hello Sean, this is James."

Sean is working out doctrines aimed at supplying livable shelter for seven billion lonelinesses around the globe. Hypothetically he and I share an acquis[39] that is quite

---

[33] "'Ah, fondest, blindest, weakest, / I am He Whom thou seekest! / Thou dravest love from thee, who dravest Me.'" The concluding lines of Francis Thompson, *The Hound of Heaven*.

[34] I am skipping over problems related to "Clasping" and the state of the cosmos prior to the cosmic change at the incarnation. Clasping has to do with the presence of the Spirit in pre-human history and gently dominating emergent probability. It is a tentative theological push on the part of McShane. See *Posthumous* 11, 12, 13, 14, 16, 19, and 21. Note that I am also skipping over problems of "Cauling" and "Calling" (see *Posthumous* 16, 17, 18, 19, and 21), which have to do with personal presence that twists through both ontic and phyletic story, in particular the calling presence of the second Persons to the first Person.

[35] "The field is *the* universe, but my horizon defines *my* universe." *Phenomenology and Logic*, *CWL* 18, 199. The italics are Lonergan's, present in his lecture notes from which this sentence is taken.

[36] *The Triune God: Systematics*, *CWL* 12, 405.

[37] *Ibid.*, 497.

[38] "Foundational talk is per se direct speech of—more precisely (about)³—fantasy and recycling." *ChrISt in History*, chapter 4, "Foundations," at page 4. "Think, for a start, of the cyclic refinement of foundations persons and of their dual task of foundational fantasy and of fostering the implementation of that fantasy. The foundational person is, so to speak, a point of pragmatic meaning, of light, in history and geography." *ChrISt in History*, chapter 2, "The General Solution to Present Ineffective Fragmentation," at page 14. *ChrISt in History* is available at: http://www.philipmcshane.org/christ-in-history.

[39] "And you can have teamwork insofar, first of all, as the fact of reciprocal dependence is understood and appreciated. Not only is that understanding required; one has to be familiar

beyond the shambles of axial "academic disciplines" and that includes a sub-word "house." We have grown slowly and sufficiently to relish "Relations" in W$_3$,[40] and we are clear that, in the concrete, a heuristics of house includes a symbolic representation of the house that Noah refurbished with Ale in mind in the film "The Notebook," as well as the five words spoken by the elderly Noah (James Garner) to his children: "Your mother is my home";[41] Bachelard's late-life house built with indomitable courage;[42] John's Jesus speaking of "many mansions" in his Father's house;[43] and Aquinas gloss on the passage.[44]

Since "being intelligent includes a grasp of hitherto unnoticed or unrealized possibilities,"[45] then a wholesome heuristics of house would also conveniently, symbolically represent future houses and housing developments that are good, where "good" is "not a system, a legal system or a moral system" but "a history, a concrete cumulative process resulting from developing human apprehension and human choices that may be good or evil."[46]

This is far too much, as is the fantasy of an elderly *Assembling, Completing, Comparing, Reduction, Classification*, and *Selection* the mangling of houses by academic disciplines better than it was, a fantasy that is presupposed by Foundations.

What I am in fact doing is an imprecise communication of the embryonic dialectic weaving of McShane around the topic of being in love with God, grappling

---

with what is call the *acquis*, what has been settled, what no one has any doubt of in the present time. You're doing a big thing when you can upset that, but you have to know where things stand at the present time, what has already been achieved, to be able to see what is new in its novelty as a consequence." *Early Works in Theological Method I, CWL* 22, 464.

[40] "Relations" is section 2 of chapter 16 of *Insight*. "3P" (Speaker, Spokener, Listener) appears above the cycling image in W$_3$.

[41] "The Notebook," 2004, starring Ryan Gossling (Noah Calhoun) and Rachel McAdams (Allie Hamilton). When Noah returns home from the war, he buys the abandoned house, fulfilling his lifelong dream to buy it for the departed Allie, whom by now he has not seen for many years.

[42] "Late in life, with indomitable courage, we continue to say that we are going to do what we have not yet done: we are going to build a house." Gaston Bachelard, *The Poetics of Space* (Boston, Beacon Press, 1969), 61.

[43] *John* 14:2.

[44] *Summa Theologica*, I-II, Q. 5, art. 2, sed contra. "*On the contrary*: 'In my Father's house there are many mansions' (*John 14:2*), which mansions, as Augustine says, 'signify varying degrees of merit in eternal life.'"

[45] *Method in Theology*, 53.

[46] *Topics in Education, CWL* 10, 33.

for a hard-won discontinuity in expression.[47] If the new science were a reality the problem of this essay would have been handed to me adequately by the group named at the end of page 250 of *Method in Theology*. In fact all I have are the leads of McShane stumbling forward from the writings of Lonergan in his usual random dialectics.

In addition I am pretending that my colleagues Terry Quinn and Patrick Brown have "precisely done"[48] their humble interpretations, and that the others on the non-existent team of cosmos-huggers have also successfully "curbed one-sided totalitarian ambitions" and precisely done and communicated his or her little part.[49]

"'The Lord is with you,'[50] and with your house."

Hypothetically, my expression about a factual and absolute cosmic lift is fine-tuned, beautiful, efficient, "purely formulated," and seamlessly received by Sean. My strange, uncommon-sense message is a precise fantasy that accelerates the others on the team of precision acrobats. The core of the handing on to Sean is a footnoteless, lean yet rich, pragmatic fantasy. Sean and I share the same faith-seeking-understanding formulae as would any collaborating scientists.

De facto neither Sean nor I are up to this type of lean conversation that is well beyond both the prose of Chardin and the lyrics of Sade. What I am doing here is a very shabby C_{5x}. How shabby?

My question goes to the heart of present pretense in Lonerganesque circles and elsewhere. It calls to mind the Epilogue of *Verbum: Word and Idea in Aquinas* regarding the identity of "understanding what Aquinas meant" and understanding "as Aquinas understood."[51] The road beyond the pretense of decadent parroting[52] is a bloody slow, careful labor of contemplation and love.[53] The road beyond shabby Trinitarian theology is an ingestion of "Relations," a sub-word of *Cherishing*.

What I can hand-on to the rest of the team, be they my colleagues in this volume or interested on-lookers, is a Foundational stand: if you or I want to get seriously to grips with 47173 and *Cherishing*, then a decent place to start is section 2 of chapter 16 of *Insight* and Appendix 3 of *The Triune God: Systematics*.

---

[47] "But here what needs luminous attention is the dangerous continuity of the conventions of expression in Trinitarian theology." Philip McShane, Q. 30 "The Trinity in History" (available at: http://www.philipmcshane.org/questions-and-answers), at page 2.
[48] *Method in Theology*, 137.
[49] See *Luke* 17:10. "So you also, when you have done everything you were told to do, should say, 'We are unworthy servants; we have only done our duty.'"
[50] See *Luke* 1:29.
[51] *Verbum: Word and Idea in Aquinas*, CWL 2, 222.
[52] "The words are repeated, but the meaning is gone." *Method in Theology*, 80.
[53] "Only by the slow, repetitious, circular labor of going over and over the data, by catching here a little insight and there another … can one hope to attain such an understanding as to hope to understand what Aquinas understood and meant." *Verbum: Word and Idea in Aquinas*, CWL 2, 223.

Foundationalizing the secondary act of existence of the incarnation is an empirical, faith-seeking-understanding self-digestion, a species of "generalized empirical method." Sub-groups of poets and mystics might inspire, but our struggle to seed global collaboration must carry with it bothersome and mostly proleptic "third way" longings and questions. Do poets and mystics contribute positively to "cumulative and progressive results"? If so, how? If not, how then?

IV. Further Contexts

In my current circumstances of semi-retirement in Mexico, now having resigned from tenure twice, in two different countries, I glimpse the madness of the academy's business and busyness. I also enjoy the leisure to slowly discover multiple personalities, little 'ol me meshed, absolutely supernaturally, with infinite We. But most of my colleagues in Latin American, the U.S., Canada, and other parts of the globe do not enjoy such circumstances or the madness of seriously asking who "I" is and who "you" is/are in the phrase: "I love you." Present circumstances of life and work simply do not offer the space and time to think in particular zones of interest, much less in the broad front of asking: "What is functional specialization?"

In this volume each person has struggled to contribute from his or her own area of interest or expertise, and each with a sense of the shabbiness of his or her contribution. Indeed, I would add that each of us has a sense of the slimness of the probabilities of successfully seeding global collaboration.[54]

What are the probabilities of getting emergent probability into the high schools in Mexico City in the next hundred years?[55] How are we to shift the probabilities of seeding global collaboration? Was that not a question that dogged Lonergan his entire life? Is it intelligent, loving, adventurous, and sane to brush aside the "third way" because it is difficult and largely a matter of heuristic adventuring on our pilgrim way to "bring consciousness in all along the line"[56] of arts and sciences, technologies, and educations from pre-school to post-graduate studies?

---

[54] The "sense" can be boosted by working patiently on the difference between summing and multiplying p's and q's and r's (probabilities). See "The Probability of Schemes" in *Insight*, *CWL* 3, 143–144. See also the conclusion of Philip McShane, "The Riverrun to God: Randomness, Statistics and Emergence," *Posthumous* 2, available at: http://www.philipmcshane.org/posthumous.

[55] See James Duffy, "El azar, la probabilidad emergente y la cosmópolis," *Revista de Filosofía* (Universidad Iberoamericana) 135: 313–337, 2013.

[56] "Either consciousness is brought in all along the line or it is not brought in at all." Bernard Lonergan, "Consciousness and the Trinity," *Philosophical and Theological Papers 1958–1964* (Toronto: University of Toronto Press, 1996), *CWL* 6, 127.

The thirteenth and fourteenth centuries missed the massive shift that Thomas Aquinas invited;[57] it would seem pretty clear by now that the twentieth and twenty first centuries are doing the same for Bernard Lonergan. Two (not) clear and broad areas of his genius—economics and the heuristics of global care—have been neglected.[58] But, the economics and the heuristics of global functional collaboration are paradigm shifts that are being resisted not merely by the general culture but by a dominant sub-group of disciples that McShane has the audacity to name "The MuzzleHim Brotherhood."[59]

What is being resisted is the full context of his reaching, which was a search for a creative grip on historical development. "It may be asked in what department of theology the historical aspect of development might be treated, and I would like to suggest that it may possess peculiar relevance to a treatise on the mystical body of Christ." [60] There is the historical development, and there is the historical development of our understanding of that development. In the past decade McShane has identified the basis of **Comparison**, named on page 250 of *Method in Theology*, as the real historical development of what he calls "The Symphony of Jesus."[61] McShane writes of **Comparison** in its fullness as the genetic ordering of historical development that gives rise to an ongoing genetic sequencing of conceptions of that symphony. The sequencing gives rise to contrafactual history and pro-factual fantasy that mediates the vision of Isaiah literally: no figure. [62] Go figure.

It is from such a non-existent genetic sequencing that I would have liked to embark on Foundationalizing *Cherishing* and thus align my will with Your will.[63] For

---

[57] See the Prologue of Philip McShane, *The Everlasting Joy of Being Human* (Vancouver: Axial Publishing, 2013).

[58] These are areas of amazing originality. The one zone that dominated Lonergan studies in the second half of the twentieth century, and that continues to dominate at present, is the zone of re-discovery, the views of minding in Aristotle and Aquinas. See further the essay cited in note 109.

[59] The title of chapter 6 of *The Everlasting Joy of Being Human*.

[60] *Insight, CWL* 3, 763.

[61] See *The Road to Religious Reality* (Vancouver: Axial Publishing, 2012), at pages 39, 42 and footnote 124 on pages 53–54.

[62] I am recalling here the concluding words of Lonergan's *Essay on Fundamental Sociology*. He quotes fully the text from *Isaiah* 2: 2-4, which speaks of the turning of swords into ploughshares and spears into sickles, concluding the essay with two powerful lines: "Is this to be taken literally? It would be fair and fine, indeed, to think it no figure."

[63] "[G]ood will wills the order of the universe, and so it will with that order's dynamic joy and zeal." *Insight, CWL* 3, 722. Cf. "For out of Jerusalem will come a remnant, / and out of Mount Zion a band of survivors. / The zeal of the Lord Almighty / will accomplish this." *Isaiah* 37:32.

the axial moment, I ride wildly in a fantasy land where "the world of interiority has been made the explicit ground of the worlds of theory and of common sense."[64]

God help me if I am posturing. Direct speech, humble howspeaking—*humilem sine fictione*—will, in good time, rebuild the house of history and her-story. It is bewildering to fathom the transposition of the "experimental issue" of *Insight* and phrases like "insight pivots between the concrete and the abstract"[65] and "Galileo had to experiment"[66] from axial blah, blah, blah land, patiently setting them into the poised flow of experimental living. Is it not?

What value do our efforts to seed functional collaboration have supposing that functional collaboration does not exist?[67] How, really, are the 747 alliance, the 47173 real relations and participations, and 2120 elimination of finance and money[68] to humanize secondary and high schools, markets and parks, bars and brothels in Morelia, the city where I live in Mexico? How might three or thirty-three little birdies pluck the phrase "some third way, then, must be found" from axial blah, blah, blah land and reggaely sing them into history?

"Don't worry about a thing,
'Cause every little thing gonna be alright.
Singing' "Don't worry about a thing,
'Cause every little thing gonna be alright!"[69]

---

[64] *Method in Theology*, 107. "To speak of the dynamic state of being in love with God pertains to the stage of meaning when the world of interiority has been made the explicit ground of the worlds of theory and of common sense."

[65] *Insight*, *CWL* 3, 30.

[66] *Ibid.*, 58.

[67] See note 9 above. A decent introduction to functional collaboration is Michael Shute, "Functional Collaboration as The Implementation of 'Lonergan's Method' Part 1: For What Problem Is Functional Collaboration the Solution?" *Divyadaan: Indian Journal of Philosophy and Education*," 24/1 (2013).

[68] Lonergan, *For a New Political Economy*, *CWL* 21, 20.

[69] Bob Marley, "Three Little Birds."

APPENDIX: "ALL WE KNOW IS SOMEHOW WITH US"

A. *Analysis: An Introduction to Proof*[70]

When the metadoctrine cited in note 23 above arrived in my inbox, I was reminded of a two-semester course, "Math Analysis," that I took at California State University Long Beach in 1986–1987. At the end of each chapter of the book, there were a dozen or more proofs on relations, functions, axioms for set theory, ordered fields, limit theorems, continuous functions, etc. I remember sitting poolside on Saturdays in Playa del Rey taking my best shot at the homework, which typically involved resolving five or six of the proofs. In the classroom on Mondays we would compare our weekend efforts.

Why was I reminded of this course? A passing score on "Math Analysis" exams was 30%. The metadoctrine is 20%–30% intelligible to me, which does not mean that I am stupid or unusually slow but that I have homework to do.

In any case, dangerous memories of plodding poolside with pencil and paper are still with me. When I saw G^i_{jk} for the first time in *Posthumous* 9: "Position, Comparison, Finite Processions," my reaction was not resentment but awe: What in the world is McShane's meaning?[71] It seems to me that the mood that is to dominate the *Novum Organon* is one of cheerful not-knowing mixed with a cherishing sharing of communicable hints.

B. Sounds of Silence

While a graduate student at Fordham University in the 1990s, I spent over two years reading Lonergan reading Aquinas[72] while reaching for an "existential ethics"[73] that

---

[70] Steven R. Lay, *Analysis: An Introduction to Proof* (New Jersey: Prentice-Hall, 1986).

[71] See also *Posthumous* 12, 14, and 19, and *Questions and Answers* 15, 21, 34, 48, and 49. They are available, respectively, at: http://www.philipmcshane.org/posthumous, and http://www.philipmcshane.org/questions-and-answers.

[72] Lonergan's doctoral thesis, "*Gratia Operans*: A Study of the Speculative Development in the Writings of St. Thomas of Aquin" (S.T.D. thesis, Gregorian University, Rome, 1940), was rewritten for publication in 1941, 1942, and 1971. The 1971 edition, *Grace and Freedom: Operative Grace in the Thought of St. Thomas Aquinas*, ed. J. Patout Burns (London: Darton, Longman & Todd), added stylistic changes as well as expanded notes to the original publication in *Theological Studies* in 1941 and 1942. "The Concept of *Verbum* in the Writings of St. Thomas Aquinas" was published between 1946 and 1949 in a series of five articles in *Theological Studies*, vol. 7 (1946): 349–392; vol. 8 (1947): 35–79, 404–444; and vol. 10 (1949): 3–40, 359–393. The articles were later published in book form in *Verbum: Word and Idea in Aquinas*, ed. David B. Burrell (London: Darton, Longman & Todd, 1967).

[73] The phrase appears in the "Questionnaire on Philosophy: Response" (1976). See *CWL* 17, 352–385. On existential ethics, *ibid.*, 357–58.

includes the precept "acknowledge your historicity."[74] To a large extent my efforts were aimed at a preliminary grasp of Aquinas' "genuine achievement"[75] as retrieved by Lonergan.

The doctoral dissertation[76] was a report on my ongoing efforts to develop a context for understanding what Lonergan might have meant by existential ethics. While working on this thesis I encountered an enormous problem—the circular labor of reading, thinking, writing, re-reading, re-thinking, and re-writing in order to understand what Lonergan might possibly have meant, while simultaneously resisting the temptation to confuse this labor with a facility with words and phrases.[77]

Acquis, the unsettling context of what had been settled by an author, was already an implicit issue for me. Besides the "spreading out, moving on, including more"[78] that occurred during Lonergan's eleven years reaching up to the mind of Aquinas, already there had been a breakthrough in economics in the early 1940s and his lifelong concern with method[79] was still years from being settled.

C. A Friendly Question

Gerald McCool, S.J., one of the readers of my dissertation, posed the question to me towards the end of the two-hour defense: "What is Lonergan up to in *Method in Theology*?" The exchange probably lasted a total of three minutes at most—I

---

[74] *Ibid.*, 378.

[75] Lonergan would later describe what he discovered in his "detailed investigations of the thought of Aquinas on *gratia operans* and on *verbum*" (*CWL* 3, 769) as "genuine achievements of the human spirit," each with "a permanence of its own." *Method in Theology*, 352.

[76] "The Ethics of Lonergan's Existential Intellectualism" is available at: http://www.lonerganresource.com/dissertations.php.

[77] "The temptation of the manual writer is to yield to the conceptualist illusion: to think that to interpret Aquinas he has merely to quote and then argue; to forget that there does exist an initial and enormous problem of developing one's understanding." *Verbum*, *CWL* 2, 223.

[78] "An Interview with Fr. Bernard Lonergan, S.J.," *A Second Collection* (London: Darton, Longman and Todd, 1974), 222–23.

[79] "I was very much attracted by one of the degrees in the London syllabus: Methodology. I felt there was absolutely no method to the philosophy I had been taught; it wasn't going anywhere." *Caring About Meaning: Patterns in the Life of Bernard Lonergan*, ed. Pierre Lambert, Charlotte Tansey, and Cathleen Going (Montreal: Thomas More Institute, 1982), 10. "The Catholic philosopher ... always tends to express his thought in the form of a demonstration by arguing that opposed views involve a contradiction. The method is sheer make-believe but to attack a method is a grand scale operation calling for a few volumes." Letter to Henry Keane, January 22, 1935. The letter may be found in the Lonergan Archives.

mentioned something about the apprehension of values in feelings or the two ways of development.[80]

In any case I remember finding it quite admirable that a respected scholar of nineteenth and twentieth century Thomism was comfortable enough—in a "friendly universe"—to confess publicly his befuddlement about *Method in Theology*. Whatever Lonergan was "up to" in this book, it did not seem to have precedence in or resonate with the works of Rousselot, Maréchal, Maritain, Gilson, or Rahner, or even, for that matter, in the earlier works of Lonergan. This is not to say that McCool was unaware of the shortcomings of what he calls a "doctrinal tradition," and that, in McCool's mind, both Rahner and Lonergan had left behind the Neo-Thomistic movement.[81] Rather it was a question about the manner or way of his leaving behind. What had happened to the classical, statistical, genetic, and dialectical methods, the notions of being and objectivity, the four chapters on metaphysics, the possibility of ethics, and the general and special transcendent knowledge of *Insight*?[82] Could it be, as Rahner himself suggested, that Lonergan had his mind on something more than method in theology, something even interdisciplinary?[83]

---

[80] See *Method in Theology*, 30–41 and Fred Crowe, "An Expansion of Lonergan's Notion of Value," *Lonergan Workshop*, vol. 7, edited by Fred Lawrence (Atlanta: Scholars Press, 1988), 35–58.

[81] See chapter 9, "The Explosion of Pluralism: The "New Theology" Crisis," *From Unity to Pluralism: The Internal Evolution of Thomism* (New York: Fordham University Press, 1989), 200–230. Unlike McCool, who was familiar with the problem of theology as it was posed by Lonergan throughout the 1940s and the 1950s, those of us who did not share that experience might wonder what the big deal was. After all, didn't Lonergan return to his usual way of thinking and writing after completing *Method in Theology*? Time and talent permitting, we have to do the library work of finding the mess in our area of interest.

[82] See the first paragraph of Philip McShane, "'What-To-Do?': The Heart of Lonergan's Ethics," *Journal of Macrodynamic Analysis* 7 (2012): 69–93.

[83] Responding to Lonergan's article, "Functional Specialties," *Gregorianum* 50 (1969), 485–505, Karl Rahner wrote: "Die theologische Methodologie Lonergan's scheint mir so generisch zu sein, daß sie eigentlich auf jede Wissenschaft paßt." "*Kritische Bemerkungen zu B. J. F. Lonergan's Aufsatz:* 'Functional Specialties in Theology,'" *Gregorianum* 51 (1971), 537. In translation: "Lonergan's theological methodology seems to me to be so generic that it actually suits every science." By the time of writing *Method in Theology*, Lonergan knew his proposal was interdisciplinary. See *Method*, 22–23, 132, and 366–67; also, in the preface to the three lectures on religious studies and theology, Lonergan explicitly states that *Method in Theology* was "conceived on interdisciplinary lines." *A Third Collection*, 113.

D. "From this sensible and palatable experiment"[84]

From 1996 until 2001 I participated in an interdisciplinary core curriculum program at Saint Mary's University of Minnesota. During the summer months, the seven or eight of us giving classes in this program found time to think about the scope and aim of interdisciplinary programs. In particular, I was asked to collaborate in the design and teaching of courses with titles such as "Our Modern Heritage" and "Great Ideas in Math and Science." In our departmental meetings we had come to an impasse regarding both the selection of texts and our understanding of what it might mean to teach texts "interdisciplinarily."

Like similar core programs, the program at Saint Mary's delegated responsibilities to individual departments—"Person" classes taught by philosophers, "God" classes taught by theologians, and "Science and Technology" taught by scientists. Because of my background in mathematics, I was called upon to teach "Great Ideas in Math and Science," which included a lab component, e.g., swinging pendulums with timing devices in hand, as well as discussions of classical texts. The experience of designing and teaching this course is an aid for interpreting "Heuristic Structures of Empirical Method," chapter 2 of *Insight*.

E. "You are the apple of my eye"

Instead of rambling down the road of a tenured professor at Saint Mary's University, I decided in the spring of 2001 to take a non-paid, no strings-attached "sabbatical" year in Mexico to learn how to teach English and to learn Spanish. The sabbatical extended to two years, then three, four, and five years, until I accepted a director position in a new bachelor's degree in Humanities and Social Sciences in 2006. The five-year sabbatical ended with the usual madness of department meetings, curriculum design, committee work, student advising, and "captación," or attracting new students, which found me "up in the air" more than I had ever been in the United States.

As a teacher of English as a Second Language (ESL), I have taught all levels, from beginners to advanced, as well as proficiency level courses on idioms, grammar, pronunciation, and vocabulary, as well as TOEFL[85] preparation.

Of the many challenges that ESL students face, one is to understand and properly use idiomatic expressions. Native English speakers do not think twice

---

[84] "From this sensible and palatable experiment it seems to me that you can very readily decide whether the reflection which comes here from the moon come like that from a mirror, or like that from a wall; that is, whether from a smooth or a rough surface." *Galileo: Dialogue Concerning the Two Chief World Systems*, trans. Stillman Drake (Berkeley: University of California Press, 1967), 73.

[85] TOEFL (Test of English as a Foreign Language) is an internationally recognized, standardized exam that purportedly measures one's competency in English.

about using such expressions as "for the time being," "head over heels," or "You are the apple of my eye," but second-language students wanting to master such expressions cling to the security of their mother tongue and spontaneously dread the "leap to a new or second identity [that] is no simple matter" and that threatens their self-identity.[86] For the adult language learner, humility is required to let down one's guard: "a context develops in which you must be willing to make a fool of yourself in the trial-and-error struggle of speaking and understanding a foreign language."[87] The advice that I give ESL students is the same advice that I have taken to heart: listen to music, read subtitles in the cinema, and pay attention to conversations on the streets, in bars, and in taxis.

More recently I have realized that I am a better ethics teacher in the ESL classroom than in the ethics classroom. In the ESL classroom, learning "survival language" is fundamental for beginning students, while authentic communication is desired by the more advanced students, who would like to move around and survive in the markets, taxis, and social gatherings; order a pizza by phone; get the young lady's or young man's telephone number at the disco; keep up with friends when they begin joking around; manage "bad words"; and communicate in English in Facebook and Twitter. These have also been my challenges as a foreigner living in Mexico.

In the ethics classroom there is no counterpart for survival language or authentic communication. By design, most ethics courses focus on *"corrientes"* (ethical currents), cases and dilemmas, "-isms," and "basic concepts" that are really just names. Aristotle's quip that "what the doctor examines is not health in this manner at all [the *Health*], but the health of man, or perhaps rather the health of an individual man, since what he cures is an individual [and not man in general]"[88] is tragicomically ignored. So much for "what is good, always is concrete."[89] What to do about the mess? Ho, ho, ho, "plainly the way out is through the more general field."[90]

---

[86] H. Douglas Brown, *Principles of Language Learning and Teaching* (White Plains, NY: Pearson Education, 2007), 70.

[87] *Ibid.*

[88] Hippocrates G. Apostle, *Aristotle's Nichomachean Ethics* (Iowa: The Peripatetic Press, 1984), 7, 1097a 11–14 (braces in original).

[89] First sentence of "The Human Good," chapter 2 of *Method in Theology*, 27.

[90] *For a New Political Economy, CWL* 21, 7.

F. "¡Aguas!"

"Agua" is Spanish for water. When expressed with emphasis it means "Be careful!" In preparation for "Functional Research," the West Coast Methods Institute 2011,[91] I spent some months asking "What does it mean to utter 'water,' not in general, but particularly?" If my water-interest is hydrostatics, then if I am becoming a GEM141[92] character, I would not treat my study of water without taking into account myself as operating subject reading or tongue-ing forth watery words while, for example, reading hydrostatic journals and perhaps even writing a non-truncated high school hydrostatics textbook; nor would I treat myself as operating subject without taking into account the object water, for example while reading articles and books in the field of hydrostatics. If my water-interest takes off from Scripture studies, then my "object" of study is the psycho-social linguistic *Sitz im Leben* of "I thirst" and "Let all who are thirsty come to the water."[93]

Could one and the same person have both water-interests, i.e., both as a physicist and as a Scripture scholar? Why not? And there is no reason to rule out personal memories of dancing waters at Disneyland, hydrodynamics, the historical emergence of the symbol "H_2O" as a breakthrough of the babbling baby in speaking about the concrete, the self-digesting repetition of this same breakthrough in oneself, and the variety of concepts and relations of a systematics of water.

> In so far as you are living a larger life, the meaning of 'water' can resonate through all ten genera of arts and a range of sciences, not to mention particular memories, enchantments, symbolisms, phobias. Water can be tongued out with a me-ning recognized by others to be elusive: it is a Ning of the poet's ownzone layer.[94]

---

[91] "The 26th Annual Fallon Memorial Lonergan Symposium in Honor of Philip McShane and His Contributions to Lonergan Studies," Loyola Marymount University, April 28–30, 2011.

[92] GEM141 refers to Bernard Lonergan, "Religious Knowledge," *A Third Collection*, 141 (top). "Generalized empirical method operates on a combination of both the data of sense and the data of consciousness: it does not treat of objects without taking into account the corresponding subject; it does not treat of the subject's operations without taking into account the corresponding object." The text has, in fact, been quoted by other authors of this volume, e.g., Henman, Mundine, and Quinn.

[93] One fine day Scripture scholars' *Sitz im Leben* will benefit from convenient symbolism W_2: $V\{W(p_i ; c_j ; b_k ; z_l ; u_m ; r_n) > HS (p_i ; c_j ; b_k ; z_l ; u_m ; r_n)$ This particular diagram appears in Phil McShane, *A Brief History of Tongue* (Axial Press, Halifax, 1998),122–23 and "Metagrams and Metaphysics," available at: http://www.philipmcshane.org/prehumous. McShane argues for the necessity of such diagramming in Phil McShane "Metaphysical Control of Meaning," *METHOD: Journal of Lonergan Studies* 24 (2006).

[94] Philip McShane, *A Brief History of Tongue*, 37.

From an appreciation of the emergence of hydrostatics and hydrodynamics as achievements of the babbling baby in history, one could go on to fantasize the emergence of macrodynamic economics as a science in the next one hundred years or so, and the effective and literal delivery of sane economics "into the hills of Ghats and onto the banks of the Godavari."[95]

G. "You are the book!"

Twice, in two consecutive semesters, the director of "*idiomas*" (languages) asked me to intervene in high school groups. Advanced students were bored, did not want to review grammar or mindlessly fill-out worksheets, and were making life extremely difficult for their teachers. On these two occasions the director has asked me to intervene by taking over the group.

After taking a day to consider the matter, I decided to intervene under two conditions. First, there would be no book; secondly, there would be no traditional final exam. The director was between a rock and a hard place and had no choice but to accept my outlandish proposal.

On Mondays I would enter the classroom and ask students to open their books to page 57 or 123. They would reply, "But James, we have no book!" to which I would reply with a smile: "That's right. You are the book!" In place of a final exam, the students modified and presented episodes of "Friends."

Nowadays in my ethics groups, after reading and/or watching Ray Bradbury's *Fahrenheit 451* with my students, I ask them to consider which book they would like to become, why that book, and how they might become it. They also act out scenes from José Emilio Pacheco's *Las batallas en el desierto*. Slowly I have been trying to aestheticize anesthetized students.

The core issue is "ordinary living that is not ordinary drama," "aesthetic liberation, artistic creativity, and the constant shifting of the dramatic setting open up vast potentialities,"[96] and the rot of "at least two centuries, through doctrines on politics, economics, education, and through further doctrines"[97] that is killing my students. If "one knows by what one is," and not by contact or confrontation, then the challenge is to become a character becoming a character capable of remembering

---

[95] A context is McShane's discussion of how Sir James Lighthill's four volumes (*Collected Papers of Sir James Lighthill*, edited by M. Yousuff Hussaini (Oxford: Oxford University Press, 1997)) carried forward the 1897 *Hydrodynamics* of Horace Lamb. "Might one not consider plausible that some dedicated Indian scientist like Lighthill in the West would bring forth a four-volume economic work in 2097 that would bring Lonergan's work effectively, and indeed literally, into the hills of Ghats and onto the banks of the Godavari." "The Global Economy and My Little Corner," *Divyadaan: Journal of Philosophy and Education*, vol. 21, no. 2 (2010), 245–256, 252.

[96] *Insight*, CWL 3, 212.

[97] *Topics in Education*, CWL 10, 232.

the future better than it was in and through "an exploration of potentialities of concrete living."[98]

H. "Teaching children children"

Two years ago I piloted a Spanish translation of *Introducing Critical Thinking*[99] in a first-year, core course on "Critical Thinking." Chapter 31, "Conversations," has been a staple in my ethics courses for six years. There are chapters on global history and feminism that help introduce in an age-appropriate way the problem of the longer cycle of decline.

There are also chapters on belief, faith, Plato's cave, *Jeremiah*, *The Bhagavad Gita*, mathematics, physics, botany, chemistry, economics, statistics, business, and history, all short chapters written not so much to provide answers as to raise basic questions. The book could be divided in two or three and used in various core courses in programs like the one at Saint Mary's University.

The huge problem that I faced two years ago was to persuade colleagues to actually self-read the little book, which they consider too easy—should not the students actually read difficult primary texts?—and/or too eccentric—the "Childout Principle"[100] sounds to them like psycho-babble. My undergraduate students find it difficult and a bit non-orthodox, but most of them appreciate the authors' efforts to reach them where they are at this point of their journey. Since the same authors hold no punches regarding mindlessness and mindfulness, in class I plead with students along the lines of Alcoholics Anonymous: "What is said here stays here."[101]

I. "What are we to do?"

In the spring of 2013, in preparation for the Second Latin American Lonergan Workshop on "The Human Good," I took some weeks to puzzle over possible ways to convince the team of organizers to try something different from the usual inefficient paper-reading and evident failure to read the first page of *Insight* and the first two pages of *Method in Theology*.[102] After compiling a list of suggestions which I

---

[98] *Ibid.*

[99] John Benton, Alessandra Drage, y Philip McShane, *Introducción al Pensamiento Crítico*, trad. James Gerard Duffy y Karla Nahmmacher de la Torre (Madrid, Plaza y Valdés, 2011).

[100] "When teaching children geometry, one is teaching children children." *Introduction to Critical Thinking* (Vancouver: Axial Publishing, 2005), i.

[101] See "Cabezas Vacías Anónimas y Comunidades Bases de "¿Huh?...¡Wow!" [Empty Heads Anonymous and Base Communities of "What?" and "Wow!"] in "La Moda y la Misión (parte 2)", available at: http://eltoquehumano-humanistas.blogspot.mx/2012_12_09_archive.html

[102] See Questions 31–37 in "Moving Lonergan Studies into Functional Talk," available at: http://www.philipmcshane.org/questions-and-answers.

had divided into *"Method in Theology"* strategies and *"Insight"* strategies,[103] I realized that they were not really two really distinct lists. In fact, I had moved strategies from one list to the other, and I could have just as easily moved them back. Is not the whole set of specialties hidden in the rewriting of high school and undergraduate texts as well as future rewritings and performances of *Insight*?

The workshop was not a big success, but there were moments of incarnate meaning that cut through the sad staleness of truncated subjects rambling on in the usual Lonerganese. In particular, a few participants decided to perform instead of deliver papers, which struck a chord with the undergraduates in attendance, who were the majority. For me the highlight of the two days was dancing salsa and swing during the lunch hour of the second day. We should have scrapped the programmed talks for the remainder of the day.

J. *"Vacaciones familiares en Acapulco: Una fábula postmoderna"*[104]

The first suggestion for changing the ethos of workshops was to tell stories. One such story that I had in mind is a story about a Mexican family planning a vacation in Acapulco—a translation and 'Mexicanizing' of the story "Toronto Family on Vacation."[105] It is a fable for teasing out the bright idea of dividing up the roles and tasks for planning the next family outing with the needs of one and all in mind. I added local flavors, customs, and idiosyncrasies to preserve the "local" in "glocal."

Reorienting the common sense of the MuzzleHim Brotherhood might be a lost cause, but our brothers and sisters are many, and many of them appreciate a good story. One day a student of mine entered the classroom and said to me, "I need therapy." I gently asked her why and she confided that the young lady writing in her journal about her macho father in the story that I had written reminded her of herself.

K. Serena Williams, Candace Perce, and Amy Winehouse

In the fall 2013 semester I piloted the first five chapters of *Futurology Express*[106] in a course called "Modernity and Postmodernity." *Futurology* replaces *metaphysics* and

---

[103] See *ibid.*, Question 34, "Some Strategies for Advancing towards Functional Specialization."

[104] "Family Vacations in Acapulco: A Postmodern Fable." This is a blog essay available at: http://eltoquehumano-humanistas.blogspot.mx/2011/11/vacaciones-familiares-en-acapulco-una.html.

[105] See Philip McShane, "A Rolling Stone Gathers *Nomos*," *Economics for Everyone: Das Jus Kapital* (Halifax: Axial Press, 1998), 149–154 and "The Turn-Around," *Futurology Express* (Vancouver: Axial Publishing, 2013), 7–13.

[106] Philip McShane, *Futurology Express* (Vancouver: Axial Publishing, 2013).

ethics as the way out of present, sick culture.[107] The first few chapters introduce the problem of collaboration; the following chapters (5–7) are an introduction to the mess in economics and the basics of sane economics. There are only five or six footnotes in the first sixteen chapters, which speaks to the problem of direct discourse.

Google Books generates a list of common names and phrases. You will find in *Futurology Express* three names that occur frequently: Serena Williams, Candace Perce, and Amy Winehouse. Serena's poise is a what-poise: "The what swayed in flexed control, standing on the globe alert to the globe up-tossed and swiped 78 feet away."[108] Candace's poise is also a what-poise about the molecules of emotion. Amy's poise is a no-poise to her truncated daddy who thinks that she is fine.

What has become clear while teaching undergraduates is that the Amys (and Andys) in my groups are better tuned to truncated, sick academic culture than are most colleagues.

L. "Assembly includes …"

In July, 2013 McShane suggested that I attempt *Assembly* as a piece of Dialectics for this volume, in particular an *Assembly* of Lonergan Studies, 1964–2013. In 2015 the project came to fruition in an essay written for a *Festschrift* in honor of Fr. Brendan Lovett.[109]

The point of gathering books, journal articles, workshops and workshop papers, courses, newsletters, interviews, websites, blogs, and video conferences was to identify integral goings-on, not a registry of persons, places, and times.[110] The challenge of the word "integral" was communicated in an email from McShane:

> The advantage of the first five years is that the theory of cyclic collaboration is missing: it appears in 1969 in the *Gregorianum* (*Method*, note page ix). Are there

---

[107] "Your daddy, present culture's idiot leadership, your daddy thinks you're fine. More automobiles and pills, work and leisure daze, more G_8s, G_{20}s etc., more unfree trade, and more \$\$\$ at the top, just bank on it!" *Futurology Express*, 107. "By replacing *metaphysics* with *futurology* in *Insight*'s definition of metaphysics I hope to rescue the neglected aspect of implementation in that definition. Also, the entire project of metaphysics is lifted into the context of Lonergan's solution to the problem of Cosmopolis, and the what-question is given its full ethical meaning." *Ibid.*, 142.

[108] *Futurology Express*, 22.

[109] "The Joy of Believing," *Himig Ugnayan*, vol. 16, *Reshaping Christian Openness*, edited by Marina Altarejos, James Duffy, and Philip McShane (Quezon City, Philippines: Institute of Formation and Religious Studies, 2016), 201–228.

[110] "If the story, the history, is to be more than just a register of places and persons and patterns in time, then one has to reach as best one can, in an emergent culture that has an ethos of such reaching, for a grip on the populations of whats going on." *Futurology Express*, 17.

bits of Lonerganism work in those years that are worth noting as anomalies, + or - ? Then there are the 45 other years, including the addition of the context of *Method in Theology* (1972): now the trail has divided: anomalies with advertence to FS-dynamic and those without.

Notice the problem of immaturity bubbling up here. James—each of you of course—has to fill in, in rough fashion, on a road not taken, but trying for an "idealized version of the past, something better than it was." *Method*, 251. There is an element here of *Completion* coming into the operation, a retro-action or cyclo-action: but let us skip past that and the manner in which it hovers over the three earlier specialties.

We need to admit the fogginess of new beginnings, as in relativity after 1904 or Quantum theory after 1926, or biology in these next decades when the word genetic stops meaning just genes carrying mythic information, codes. We are helped here by imagining parallels to successful sciences, where a standard model is in decent control and we are fine-tuning real situations. At present we are monstrously far from that in futurology or in its precursors theology and philosophy.

So: a stumbling selection of the really bad and really promising in the past fifty years of Lonerganism. Pat Brown has that great instance of the really bad: the dodging or misinterpretation of *Method* 250.[111]

The *Method* 250 prescription states that *Assembly* and the other five italicized words will be performed differently by different investigators and that the differences "will be brought out into the open." I decided to write this appendix in the spirit of bringing James out into the open. Such bringing out one day will be beautifully and brutally central to Dialectics as science, to the concomitant interdependence of Dialects and Foundations, and to the indirect interdependence of the first and second phases.[112]

---

[111] I quote from an email that Phil McShane sent to William J. Zanardi and James Gerard Duffy, July 31, 2013, with the title: "trying dialectic analysis." The "dodging or misrepresentation of *Method* 250" is a reference to Patrick Brown's "Functional Collaboration and the Development of *Method in Theology*, Page 250," *Reshaping Christian Openness*, 171–200.

[112] See *Method in Theology*, 144.

6. THE FIFTH FUNCTIONAL SPECIALTY AND FOUNDATIONS FOR CORPORATE LAW AND GOVERNANCE POLICIES

Bruce Anderson

I. Context

Introduction

When I graduated from university in 1978 I didn't understand why I couldn't get a job. Why was the unemployment rate in Nova Scotia 9%? In 1983 when I graduated again, I couldn't understand why I had to pay 18.5% interest on my student loan. In 1992, when I was living in England, it baffled me why George Soros was not jailed for shorting the pound, forcing the UK Treasury to spend billions to prop up the pound, and ultimately resulting in the UK leaving the European Exchange Rate Mechanism. And in 2008 I shook my head at the massive government bailouts of companies that took huge risks that were considered "too big to fail." Today I wonder why so many Ph.D. students in accounting, finance, and economics are obsessed with correlating executive compensation and share prices.

Since 2001 I have been teaching commercial law to students in a business school. During the past ten years business ethics has turned from an occasional topic of conversation in the photocopy room to a mandatory course for all undergraduate business students. I still find most of the issues and debates quite banal. Being an ethical business person simply amounts to avoiding conflicts of interest and not doing things that are dishonest or illegal. One chunk of my commercial law course is about the different ways businesses are organized. I cover the legal aspects of sole proprietorships, general partnerships, limited partnerships, limited liability partnerships, private and public corporation, income trusts, and co-operatives. I also point out the competing views of the shareholder primacy model of corporate governance and the stakeholder theory. In light of what I know about Bernard Lonergan's five-square diagram, and what I know about co-operatives, those topics have also failed to grab my attention. I even wrote a paper called *Is There Anything Special about Business Ethics?*[1] My answer there was that for business people to act ethically they have to know how an economy works, is working, and to respond appropriately. I drew on the two years of full-time work comparing and contrasting the views of Gregory Mankiw with the economic science of Bernard Lonergan. Hence any talk of business ethics by someone who does not know something about criminal law, corporate law, and securities law and does not have a grip on Lonergan's five-square diagram I now judge to be simple-minded. It was

---

[1] Bruce Anderson, "Is There Anything Special about Business Ethics?" *Journal of Macrodynamic Analysis*, vol. 7 (2012), 54–68. This article is available at: http://journals.library.mun.ca/ojs/index.php/jmda/article/view/360/232

obvious to me that business people charged with fraud and insider trading and others who played key roles in the 2008 global financial crisis were greedy. Much less obvious is the role of deranged economic theories in corporate decision-making.

Despite my cynicism I was intrigued by a paper written by Michael Marin called *Disembedding Corporate Governance: The Crisis of Shareholder Primacy in the UK and Canada*. In that paper he convincingly argued that "shareholder primacy creates perverse corporate incentives that are detached or disembodied from their economic, political, and social context."[2]

The shareholder primacy model of corporate law and governance can be summed up in the following way. The directors and officers of a corporation have a legal duty to act in the best interests of the corporation. This legal duty has being interpreted by the courts to mean that directors and officers are to place the interests of shareholders before all other stakeholders. In practice this means that directors and officers have a legal obligation to maximize the value of shares held by shareholders. Shareholders, then, are to be the sole beneficiaries of corporate business activities.

Marin's claim is that "these incentives [to maximize shareholder value] were the root cause of the global financial crisis." More broadly, he argues that

> the trajectory of the global financial crisis in Canada and the UK was primarily caused by the relative disembeddedness of large financial institutions. In particular, excessive risk taking by those institutions, combined with a weakening of prudential regulation, led to the emergence of a system divorced from its economic and social context.[3]

Of course, I had read, and even taught, Joel Bakan's views on the psychopathology of large corporations,[4] but Michael Marin seemed to be on to something. Marin also mentioned a recent paper by Allan Hutchinson who had written a book ten years ago arguing that corporate governance should be more democratic. Hutchinson's new paper begins confidently with the assertion that "the whole nature of what counts as good corporate governance must be rethought and reconstructed from the ground up."[5] My suspicion was that I was really onto something, some serious attention had been paid to the problem of how companies are structured and governed. Business ethics, surprisingly, had become interesting!

---

[2] Michael Marin, "Disembedding Corporate Governance: The Crisis of Shareholder Primacy in the UK and Canada," 39/1 *Queens's Law Journal* (2013), 223.

[3] *Ibid.*, 223.

[4] Joel Bakan, *The Corporation: The Pathological Pursuit of Profit and Power* (New York: Penguin, 2009).

[5] Allan Hutchinson, "Hurly-Berle: Corporate Governance, Commercial Profits, and Democratic Deficits," 34/4 *Seattle University Law Review* (2011), 1219.

This essay falls into the four parts that were suggested as a general structure for the contributions to this volume. The key part is, of course, the second part which identifies a foundational advance in the interplay of economics and law. The other parts, however, are essential to placing that identification in a realistic context that has all the initial appearance of being a surrealism. We are, in this volume mapping a future that, according to Philip McShane, may shape up to a global sanity in the Tenth Millennium. Here I am reaching for foundations of a quite new economics and a dynamic of policy-formation in law that, to say the least, is not in the usual mold.

Haphazard Dialectics, Foundations, and Policy

Before immediately proceeding to the focus of the paper, namely FS5/Foundations, it is worth identifying aspects of dialectics, foundations, and policy in the papers written by Marin and Hutchinson mentioned above. I begin by identifying elements in Marin's paper that can be cast as a swing from critical analysis through foundations to policy. Of course, these aspects are not explicit and remain undifferentiated in his work. The same can be said for Hutchinson.

Michael Marin's *evaluation* of the shareholder primacy model of corporate law and governance takes two tacks. One,

> the economic instability during the global financial crisis provides ample evidence that the incentive structure of shareholder primacy is destabilizing to financial institutions. In particular, the corporate governance laws of the UK and Canada encourage and arguably require these institutions to maximize profit, which leads to the excessive reliance on financial innovation in order to boost leverage.[6]

In other words, the current legal framework of corporate governance reinforces excessive risk taking. Two, "it is increasingly evident that a corporate governance regime which treats social and political outcomes as an abstraction is both unrealistic and destructive."[7]

Marin wants to "replace shareholder primacy with an embedded model of corporate governance that reflects the economic, political, and social influence of financial institutions."[8] This is the *ground* or *foundation* of Marin's solution to the shareholder primacy problem.

He offers two ways to replace the shareholder primacy model. In doing so, Marin moves from his critical comments on the shareholder primacy model, comes up with the foundations for a solution mentioned immediately above, and then swings forward to his two general lines of solution, what might be called *policies*. One

---

[6] Marin, 259.
[7] *Ibid.*, 268
[8] *Ibid.*, 269.

option is to cut shareholders out as the primary figures in corporate governance in a similar way that credit unions and building societies are governed by members/customers. Option two is to make directors, officers, and shareholders liable for adverse economic and social outcomes.

Let's turn to Allan Hutchinson. His *critical judgment* is that shareholder primacy as the preferred rational for corporate law and governance promotes 'short-termism' and 'executive self-interest' that lead to the risk taking of the global financial crisis.[9] To be more precise, for him the problem is that corporate governance has become "fixated on inflating or maintaining share price in the short-term."[10] Further, he believes that "traditional theorizing has failed to make a persuasive case for how the modern corporation can be reconciled with the rhetoric and reality of democratic governance in contemporary society."[11] To state it simply, corporate governance neither embodies nor supports democratic values.

Hutchinson proposes to replace the shareholder primacy model. Part of the *ground* or *foundation* of his proposal is his insight that

> the corporate form is a distinctly public institution that is brought into existence by the state and has certain conditional powers delegated to it by the state. As constructions and emanations of the state, modern corporations have a distinctly public origin and a decidedly public purpose. The debate about corporate governance is, therefore, about the nature of parameters of public purposes.[12]

The key issue is "what public interests should the corporation pursue and how it should go about formulating and operationalizing them."[13] The ground of his proposal includes his arguments supporting a democratic agenda for reforming corporate governance. His goal is "to ensure that large corporations act in a more democratic and responsive manner."[14] "If capitalism is to remain, then it must serve rather than master the interest of democracy."[15]

Allan Hutchinson divides his reforms into four groups. These are his *policies*. First, he argues that the limited liability of shareholders should be modified to better balance control and risk in that if you have more control you should bear more risk. Hence selective liability should be imposed on controlling shareholders, the limited

[9] Hutchinson, 1221.
[10] *Ibid.*, 1221.
[11] *Ibid.*, 1248.
[12] *Ibid.*, 1249.
[13] *Ibid.*
[14] *Ibid.*, 1285.
[15] *Ibid.*, 1257.

liability of shareholding corporations should be abolished, and the vicarious liability of directors in some situations should be greater.[16]

Second, he believes that the fiduciary duties of directors should be broadened so that corporate actors and its beneficiaries shoulder more responsibility. To that end, a more comprehensive and creative understanding of "in the best interests of the corporation" should be developed that is more democratic and public in its orientation, and not reducible to a simple formula, or set of fixed interests, such as maximizing shareholder value.[17]

Third, Hutchinson advocates that boards of directors include "stakeholder groups whose interests are directly and substantially at stake in corporate behaviour."[18] He singles out three groups that should be represented: shareholders, employees, and other stakeholders or the public such as creditors, suppliers, customers, and the local community.

Fourth, he thinks there should be a "public regulatory body whose exclusive responsibility is to deal squarely with corporate governance."[19] Its aim would be to represent the public interest. Also, minimum regulatory standards should be enacted to ensure corporations do not act against the public interest, are more transparent in their transactions and dealings, disclose their actions regarding economic and social issues, product safety, compliance structures, labour practices, environmental effects, charitable donations, and political contributions.[20]

A few comments are in order. These proposed reforms, like those of Marin, can be seen as a move from an evaluation of the shareholder primacy model, and depend on what I have identified as efforts to articulate the ground or foundations for new creative policies. Both Marin and Hutchinson break from criticism of shareholder primacy and swing to new policies for corporate law and governance by searching out the ground or foundation that will enable them "to go beyond the structural tinkering that characterizes present reform efforts."[21]

I accept their criticisms of shareholder primacy as valid. The reforms on offer strike me as good practical solutions to problems that have been proposed by individuals familiar with a broad range of issues related to corporate law and governance. Nonetheless, however much the policies proposed by Marin and Hutchinson make sense in light of their progress from criticism of the shareholder primacy mode of corporate law and governance, to articulating the ground for new policies, and then to proposing new policies, something is still missing.

---

[16] *Ibid.*, 1251.
[17] *Ibid.*, 1252.
[18] *Ibid.*, 1253.
[19] *Ibid.*, 1256.
[20] *Ibid.*
[21] *Ibid.*, 1223.

Consider Marin who wants a system of corporate law and governance which is not detached, disembodied, or divorced from its economic, political, and social context.[22] This is what he identifies as the ground of new creative policies. But, for instance, which aspects of the economic context should be taken into account and ground policies? The current economic context can be summed up in terms of Gregory Mankiw's *Ten Principles of Economics*: (1) People face trade-offs; (2) The cost of something is what you give up to get it; (3) Rational people think at the margin; (4) People respond to incentives; (5) Trade can make everyone better off; (6) Markets are usually a good way to organize economic activity; (7) Governments can sometimes improve market outcomes; (8) A country's standard of living depends on its ability to produce goods and services; (9) Prices rise when the government prints too much money; (10) Society faces a short-run trade-off between inflation and unem-ployment.[23] Should these ten principles constitute the economic context? Philip McShane and I wrote a book showing that these principles are either trivial or mistaken.[24] Our current economic context also comprises ever-increasing income inequality, poverty, boom and bust cycles, and money has turned into a commodity. Our lives our dominated by the gamblers on Wall Street and The City, finance is detached from the productive process, profit maximization is the aim, and self-interest the dominant value. Wouldn't we, in fact, be better off if new policies and laws concerning corporate governance were detached, disembedded, and divorced from our current economic context?

Now let's try to identify what is missing in Hutchinson's analysis. The ground of Hutchinson's reforms includes his declaration that capitalism must serve democracy.[25] But what does he mean by 'capitalism'? What exactly does he mean by 'serve'? Does he mean by 'capitalism' the deranged and destructive economic structures, activities, and culture that is at the heart of our economic systems which I listed in the previous paragraph? How can that ever be a suitable ground for more democratic corporate governance? Perhaps you are also beginning to suspect that something is missing. That something is missing can be captured by the following two questions: Is there such a thing as sane economics? If there is, wouldn't we want sane economics to ground our policies on corporate law and governance?

Evidently, there is a gap in the foundation of Marin's and Hutchinson's policies, and you would also suspect that this hinders the development and formation of policy that really gets to the root of corporate law and governance.

---

[22] Marin, 223.

[23] Gregory Mankiw, *Principles of Macroeconomics* (Montreal: The Dryden Press, 1997).

[24] Bruce Anderson and Philip McShane, *Beyond Establishment Economics: No Thank-you Mankiw* (Halifax: Axial Press, 2002).

[25] Hutchinson, 1257.

What is absent is "the missing understanding at the heart of all political, financial, and business doings."[26]

II. Content of FS5

The context for my reach here is my foundational perspective, which can be identified descriptively with the listing in *Method in Theology* on pages 286 to 287. But the content of my extension of foundations that I would hand on to the sixth functional specialty is a precision of meta-policies regarding legal creativity in the face of fundamentally destructive elements of financial operations that have become accepted and conventional in the past sixty years. That destructiveness is identified properly only within a new scientific economics that accounts for the function of finance in global progress. Hence the key question is: *What is sane economics?*

What I am searching for here are foundations for policies concerning corporate governance. My guiding question is "What is the ground for doctrines about how to intelligently govern and manage corporations?" It seems best to begin by offering a simple sketch of aspects of that ground.

Business people have to know how an economy works, is working, and how to act intelligently in light of that knowledge. The key is getting to grips with Bernard Lonergan's five-square diagram and appreciating that the governance of corporations is governed by the oscillations of local, national, and global flows of production and finance. The ground of good corporate governance requires that those running our corporations understand how the production and sale of goods and services is connected to the circulation of money in a properly functioning economy. Further, they must understand what the needs and demands of such a system are, and be able to do what is necessary to keep the system functioning properly. In other words, the foundation of good corporate governance is knowledge—how an economy works, and is working—and taking the responsibility to act intelligently—that is to keep the productive and financial rhythms in step in order to provide the goods and services people need.

To give some idea of the most basic knowledge of economics necessary to ground good corporate governance it is worth quoting a very compact version[27] of Lonergan's explanation of how the parts of an economy fit together:

- Goods and services should be divided into two distinct classes: capital goods and services and consumer goods and services.

---

[26] Philip McShane, *Futurology Express* (Vancouver: Axial Publishing, 2013), 103.
[27] Bruce Anderson and Philip McShane, "Grounding Behaviour in Law and Economics," *Legislation in Context: Essays in Legisprudence*, ed. Luc Wintgens and Phillippe Thion, Ashgate, Surrey, Chapter 8, 157-169, 163.

- The productive rhythm of capital goods and services is related to a particular circuit of money consisting of expenditure, receipts, outlay, and income. The productive rhythm of consumer goods and services is related to its own particular circuit of money consisting of expenditures, receipts, outlays, and income.

- The velocity of money in these two main monetary circuits is tied to the velocity with which goods and services are produced and sold.

- In redistributive exchanges the velocity of money depends simply on how often owners buy and sell things, not on the production of goods and services. For instance, shares in a corporation can be sold whenever a seller has a buyer. Such sales are not part of the productive process.

- When an economy is running properly the production of goods and services is subject to phases. They are capital expansion, consumer expansion, and steady-state.

- During a capital expansion the economy devotes its resources to increasing the quantity of capital goods and services.

- When the production of capital goods and services reaches its maximum, the resources of the economy should then be devoted to a consumer expansion—increasing the production and sale of consumer goods and services.

- When the production of consumer goods and services reaches its maximum, the economy experiences a steady-state phase during which neither the production of capital nor consumer goods and services are increasing or decreasing.

- Raising the standard of living of citizens depends on realizing the basic expansion phase of the economic cycle.

Such knowledge leads to formulating norms. Good corporate governance would have to obey the following norms of a properly functioning economy. To state it another way, knowledge of a properly functioning economy calls for responsible policies in order to meet the needs of the economy. Again, a compact list will have to do:

- Money has to keep pace with the production of goods and services. For instance, in order to increase the production of capital goods there must be a proportionate increase in money used for the production and sale of those capital goods.

- The cross-overs must be balanced. One monetary circuit cannot expand at the expense of the other. For instance, if money is removed from the consumer circuit by taxation and spent on research and development there will be less money available to pay for the production of consumer goods and less money available to buy them. Money removed by taxation from either circuit must be returned to that circuit or else that circuit will shrink.

- Money leaving an economy to pay for imports must be replaced or else the economy will shrink.

Intelligent and responsible actions are called for on particular occasions in order to ensure that an economy is not destabilized. For example:

- When an economy is in a steady-state phase, do not add money to it.

- Artificially promoting a capital expansion should not be permitted.

- During a capital expansion do not raise the wages of workers.

- Support the shift from a capital expansion to a consumer expansion.

- Transfer payments from one part of a country to another part, for instance, should not be permitted unless it is known they will not adversely affect the circuits.

- Money is a system of public book-keeping; it should not be treated as a commodity, but as a form of promise.

- Currency speculation should not be permitted.

- Credit should be given to people in order to finance good ideas. 'Loaning people all up,' easy money policy, cheap loans, slacker lending standards, risk-spreading, and securitization should be forbidden.

Perhaps it is becoming evident that good corporate governance is a particular zone of sane economics. The foundation of good corporate governance is knowledge of economic science which, in turn, calls for suitable actions at appropriate times, and other actions that should not be permitted. With this new context of economic science we notably left the zone of common sense street policies and entered the zone of sane economic theory grounding democracy.

Of course, it is impossible to legislate the proper actions for all occasions where good corporate governance is demanded. However, unacceptable behaviour can be identified and perhaps spelling it out and assigning legal sanctions would be useful. For instance, there could be a law fining or jailing directors and officers whose corporations destabilize the monetary circuits by, for instance, speculating on currency rates. Laws prohibiting directors and officers from artificially maintaining a capital expansion may be appropriate. Laws prohibiting and punishing directors and officers from increasing wages during a capital expansion might work. Punishing

123

directors and officers of companies who blocked the shift from a capital expansion to a consumer expansion might be appropriate. Laws punishing foreign companies and their officers and directors from financing a domestic surplus expansion may prove necessary. And it might be worthwhile making it a criminal offence for corporations and their directors and officers to treat money as a commodity.

III. Handing-On from FS5 to FS6

In the previous section I pinned down the foundational insight that *corporate governance and law must be grounded in sane economics.* I also listed various aspects of scientific economics. The topic in this section is how I hand-on that insight to policy-makers, FS6. Of course, the handing-on here is illusory in that there is no group committedly lifting a theoretically sound economics into the context of law-forming policies. Indeed, present legal practice and theory fail to distinguish in any serious way policies that are expressed by common sense from policies that are grounded on the discussion in the previous section.

But let's assume there is a group specializing in FS6. Broadly speaking, their job is to articulate very precisely understood guidelines, doctrines, policies. But more specifically, in terms of corporate governance, their task is to come up with a replacement for the shareholder primacy model. The key contribution of FS5 to the efforts of FS6 is the foundational insight that FS6 must ensure that the new model of corporate governance and law they create meshes with economic science, including the norms for a properly functioning economy I listed in the previous section.

But what else might be useful to hand-on? The various legislative actions I identified at the end of previous section might be helpful to FS6 when they tackle the job of drawing the line between acceptable and unacceptable corporate behaviour.

Also, it is important to stress to FS6, and no doubt taken for granted by FS6, that the replacement of the shareholder primacy model must also fit with other fundamental economic doctrines. For instance, the replacement must be consistent with the following doctrine on profit expressed by Philip McShane:

> Profits are of two kinds. There is ordinary profit that is required in order to ensure maintenance of the economic enterprise. Then there is the profit that can enable the implementation of an innovative idea within the business. The latter profit may not be available in coincidence with, concomitant to, the new idea, in which case there is the financial facility of credit. That facility, in the presence of a new idea, is regularly creative of what is called money.[28]

---

[28] Philip McShane, *Futurology Express* (Vancouver: Axial Publishing, 2013), 83.

To state it another way, the replacement for the shareholder primacy model of corporate law and governance must be consistent with the stand on money and profit that emerges from understanding the five-square diagram.

However, the task of replacing the shareholder primacy model will not be easy for FS6. Perhaps a statement of Lonergan best captures the extent of the challenge.

> Now to work out in detail the conditions under which this must be done, and to prescribe the rules that must be observed in doing it, is a vast task. It means thinking out afresh our ideas of markets, prices, international trade, investment, return on capital. Above all it means thinking out afresh our ideas on economic directives and controls. And if we are to do this, not on the facile model of the totalitarian or socialist regimes which simply seek to abolish the problems and with them human liberty, then there will be need not merely for sober and balanced speculation but also for all the concrete inventiveness, all the capacity for discovery and for adaptation, that we can command.[29]

IV. Handing-on from FS5 to C9

I want to hand-on the same foundational insight—corporate governance and law must be grounded on sane economics—to C9 specialists, experts in communicating the discoveries of functional specialists to non-specialists. At the present time, this type of hand-on is necessary because there is no specialized FS6 group to accept my insight and take on the job of creating a new model of corporate governance. Hence, in order to replace the shareholder primacy model with a sensible alternative, the findings of FS5 need to be communicated to interested non-functional specialists with the hope of seeding further inquiry and change.

During my 'hand-on to FS6' I was not only concerned with calling attention to my foundational insight, but also pointing out complementary norms and economic doctrines that are relevant to the job of creating a new model of corporate governance. My expression there was directly to them. I said exactly what I meant and I took it for granted that my audience, FS6, had a decent grasp of economic science and a rich foundational perspective akin to that described in *Method in Theology* on pages 286 and 287. But the audience here is very different. My audience is not thus cultured, and the role of C9 here is to communicate my foundational insight to people who are not engaged in FS1 to FS8. So, my question is how should I hand-on my foundational insight to people lacking in the needed culture? What can I hand-on that would help them do their job?

To tackle those questions let's return to Michael Marin and Allan Hutchinson, two non-functional specialist scholars with a deep interest in corporate governance. What is the best way for C9 experts to communicate to them the idea that a new

---

[29] Bernard Lonergan, *For a New Political Economy*, *CWL* 21, 105–106.

model of corporate governance and law must be grounded on sane economics? Further, what is the best way to get across to them the fact that the policies and doctrines of FS6 are quite remote from their own proposals? Their horizon is that of common sense. They offer immediate practical proposals to improve corporate governance, but they are not grounded on understanding how an economy works, is working, and what needs to be done to keep it working. The five-square diagram, for instance, is beyond their horizon. The aim, then, of C9 is more modest than that of FS6. The immediate aim of C9 is to communicate my foundational insight, not to get to work replacing the shareholder primacy model.

Back to what I should hand-on to C9. What do I think would help C9 talk to people like Marin and Hutchinson? I can identify two things. One, I want to stress that C9 must have a decent grasp of economic science. I would say they have to be at home with the discussion in the *Content of FS5* section of this paper. Also, they have to understand the stand on money and profit expressed by the fundamental doctrine on profit, which I quoted above, which emerges from understanding the five-square diagram. In fact, without a clear view on how an economy works, legal and policy manoeuvrings about corporate governance, tax, trade, equality, democracy, progress, culture, values, whatever, are manoeuvrings in the economic dark. And replacing the shareholder primacy model of corporate law and policy depends on understanding how any economic unit works, and is working. A crucial part of that clear view is that there are two main sets or flows of payments in an economy—capital and consumer—and that without making that distinction corporate governance is in the field of guess-work and myth. Any legislation and policies regarding corporate governance that fail to take that distinction into account will be haphazard and arbitrary. Policies and laws regarding good corporate governance, then, depend on the measure of one's understanding of the normative demands of the economic process, and one's understanding of the activities of particular corporations in particular economies. Policies and legislation regarding corporate governance and law should be based on such understanding. However, the point I want to communicate to C9 is that without such knowledge C9 will be unable to do its job.

Two, I believe it might be worth stressing to C9 that the replacement for the shareholder primacy model of corporate governance worked out by FS6, and grounded on sound economic science, will be quite remote from the common sense proposals presented by Marin and Hutchinson promoting economic stability, personal responsibility, democratic values, and the public interest. A possible communication strategy for C9 is to try to get across the idea that something is missing from Marin and Hutchinson's analyses, namely that corporate governance must be grounded on sane economics. Then, of course, the challenge is to communicate what is meant by sound theoretical economics. Perhaps in attempting to do that, it will dawn on C9's audience that Marin's call for a "model [of corporate governance] that reflects the economic, political, and social influence of financial institutions"[30] is actually a gesture without content and context. And it might slowly dawn on C9's audience that Hutchinson's plea for capitalism to support democracy, and inserting greater personal responsibility of directors, democratic values and the public interest into corporate governance, are rough and ready proposals compared to knowing the dynamic nature of economies—their two oscillating productive circuits, redistributive flows, cross-over flows, economic cycles—and the norms businesses must respect to keep the economic show on the road. My point here is simply that it might be useful to remind C9 how big a difference there is between common sense, immediately practical solutions to problems and theoretical understanding, and the difference between half-measures and long term solutions so they can that get across to their audiences.

V. Concluding Context

My suggestions are made, then, in a general methodological vacuum. My colleague, Michael Shute, in his essay in this volume, faces that vacuum in a creative pragmatics of novel popular communications. That essay is, like my own and Philip McShane's essay, foundational. McShane pushes for a re-conception of heuristic control of situations. And Shute reaches for a new perspective, within that heuristic, on actual effective interventions in formal and informal economic education. In this essay, I have reached for a fuller perspective of a massive shift of policies regarding corporate governance and law, a shift that would require the initial realization of suggestions made by Shute and McShane. But the initial realization of Shute's suggestions are concrete possibilities of the near future that, as he points out, is of the essence of statistically effective implementation.

I end by musing over the aftermath of that successful realization, the hope for a groundswell of objection to present destructive economic and legal practices that would shake the fixity of these disciplines, condemn current financial practices, and ground the emergence of a legal tradition that can delineate patterns of business and economic destruction.

---

[30] Marin, 269.

7. HOUSING POLICIES IN AND ON THE MODE OF THE SIXTH FUNCTIONAL SPECIALTY

Sean McNelis

At a first approximation, one thinks of the course of social change as a succession of insights, courses of action, changed situations, and fresh insights. At each turn of the wheel, one has to distinguish between fresh insights that are mere bright ideas of no practical moment and, on the other hand, the fresh insights that squarely meet the demands of the concrete situation. Group bias, however, calls for a further distinction. Truly practical insights have to be divided into operative and inoperative; both satisfy the criteria of practical intelligence; but the operative insights alone go into effect for they alone either meet with no group resistance or else find favor with groups powerful enough to overcome what resistance there is.[1]

Introduction

In this paper in and on the sixth functional specialty Policies, I will use housing as a case study. The paper proposes some housing research policies and some housing policies and, shows how in operating in Policies, a researcher will draw upon the results of Foundations and pass on their results to Systematics.

I have divided the paper into five parts each with a number of sections. Part Ia begins by outlining three contexts for this paper: a personal context; the context of my struggle towards understanding and then operating within the framework of Functional Collaboration; and, the context of presuppositions of doing 'housing policies' and the consequent personal and collaborative difficulties in doing 'housing policies' in the absent context of functional collaboration in housing.

Part Ib is prior to Policies and seeks to fill in, in a common sense way, the gaps of the first five functional specialties. Part II seeks to operate with the functional specialty Policies and propose some policies for housing research and housing, while Part III outlines the task I am handing on to the seventh functional specialty, Systematics and Part IV reflects on what I have done.

Here, I must also make two introductory notes of clarification. First, I note that Lonergan in *Method of Theology* uses the singular when referring to functional specialties that relate to understanding the past—Research, Interpretation, History, and Dialectic. But he uses the plural in relation to functional specialties looking to the future—Foundations, Doctrines, Systematics, and Communications. This seems to me to indicate the plurality of expressions of the human spirit. So, in my discussion of Housing Policies I will pick up various expressions of housing.

---

[1] *Insight, CWL* 3, 249.

Second, throughout the essay I use the term 'Policies' rather than 'Doctrines.' I do this for two reasons. As outlined in the personal context in Part 1a, throughout my working career I have been involved in proposing and advocating housing policy and this is the term most commonly used in everyday discourse and in the social sciences. For me, the term 'policies' is associated with a wider meaning of action and new directions whereas 'doctrines' is associated with statements and verbal expressions.

Ia. Contexts

A personal context

I first encountered Lonergan's *Insight* and *Method in Theology* in the early 1970s. After having taken *Method in Theology* as a subject within a theology degree in the late 1970s, my concerns and focus shifted into housing policy.

For over thirty-five years, I have been involved in housing management, housing policy and housing research. For two years I worked as a housing manager living on a high-rise public housing estate in Melbourne, Australia. I later went on to become a founding member, director (chairperson/treasurer) and tenant of a small housing co-operative in inner Melbourne. As a community development worker for five years, I was involved in the development of different types of housing organizations—emergency housing services, housing information and referral services, housing co-operatives and tenant organizations. Over 12 years as a housing policy worker I was in the forefront of community organizations researching, analyzing and critiquing government housing policies, advocating for better housing policies and developing infrastructure for the newly emerging community housing sector in Australia.

Over the past 12 years, I have shifted more into housing research rather than housing policy at the Swinburne Institute of Social Research (within Swinburne University of Technology).

During my time as a housing policy worker and research, I wrote (or was involved in a team that wrote) papers on a range of topics including:

- various submissions on Australian and Victorian[2] housing policies;

- papers envisaging a future for social housing, reviewing performance monitoring, discussing the future of high-rise public housing, advocating housing affordability for low-income households and proposing new housing directions for people with a disability;

- Australian Housing and Urban Research Institute reports on a private investment vehicle for community housing, on independent living units

---

[2] Victoria is one of the seven states in the Australian federation.

for older people with low incomes and low assets, on rental systems, on older persons in public housing, on the motivations of private rental investors and on asset management; and,

- local government housing strategies.[3]

Housing policy and research in which I was involved is deeply steeped in the world of common sense.

The personal struggle towards this new science of progress

Soon after coming to the Swinburne Institute for Social Research in 2002, I found myself confronted by a vast array of different types of housing research and by the many debates (at times acrimonious) among researchers. It seemed to me, however, that these debates were often at cross-purposes because the researchers were doing different things. In reaching this conclusion, I was recalling Lonergan's work on functional specialization that I had encountered 30 years ago.

This set me on the path of further exploration of Lonergan's writings, in particular functional specialization.

As I began more intensive work on functional specialization some ten years ago, Lonergan pointed me to something beyond my current understanding of science. He proffered an invitation. My beginning was simply a matter of belief[4] that concerted work on Lonergan's writings would take me into a new world, open up new vistas and new horizons. He, like many scientists, writers, visionaries and social activists, is a prophet pointing to something better. He, like many others, invites us: to become better persons; to take up new challenges; to become cognizant of the injustices we perpetrate and feel the pain of disappointments and sorrows in those around us; to stop and enjoy the moments of laughter and of achievement of ourselves and those around us; to ask new questions and seek new answers; and to love more deeply and intimately. Only after ten years solid work as I 'retrace my steps' can I appreciate that I have reached a point where I'm beginning to understand something of what he is offering.

I began by re-reading *Insight* and *Method in Theology* as well as Ken Melchin's *History, Ethics, and Emergent Probability*.[5] This expanded into the writings of Philip

---

[3] For references to these various papers/reports, see the bibliography in Sean McNelis, *Making Progress in Housing: A Framework for Collaborative Research* (London: Routledge, 2014), 232–255.

[4] See *Insight*, CWL 3, ch. 20, section 4, 725–40.

[5] Kenneth Melchin, *History, Ethics, and Emergent Probability: Ethics, Society, and History in the Work of Bernard Lonergan* (2nd ed.)(Canada: The Lonergan Website, 1999).

McShane,[6] James Sauer,[7] Matthew Lamb,[8] and William Mathews,[9] complemented by further writings of Lonergan (in particular his two volumes on economics)[10] and Melchin.[11]

I was already dissatisfied with the way in which housing research and policy was conducted and its lack of traction within government decisions. My reading began to raise fundamental questions about the current culture of housing research. I became more and more critical of this culture and became convinced that it needed radical transformation if it was to provide practical and innovative advice to decision-makers.

In my reading, I recall four inter-related 'break-through' moments. The first was the discovery that an explanation of something (such as housing) grasped the 'functional relations' between the relevant, significant, and essential elements that constituted this something. This grasp left aside those elements that were irrelevant, insignificant, and incidental.[12] This understanding contrasted markedly with most social science research, particularly economic research that I had encountered. Social

---

[6] See the collection of published books and website series by McShane available at: http://www.philipmcshane.org.
[7] J.B. Sauer, "Economics and Ethics: Foundations for a Transdisciplinary Dialogue," *Humanomics*, vol. 11, no.1/2 (1995), 5–91; Sauer, "Meaning, Method, and Social Science: A Realist Account," *Humanomics*, vol. 18, no.3 (2002), 101–113; Sauer, "Metaphysics and Economy—the Problem of Interest: A Comparison of the Practice and Ethics of Interest in Islamic and Christian Cultures," *International Journal of Social Economics*, vol. 29, no.1/2 (2002), 97–118; Sauer, "Christian Faith, Economy, and Economics: What Do Christian Ethics Contribute to Understanding Economies?," *Faith and Economics*, vol. 42, Fall (2003), 17–25.
[8] Mathew Lamb, *Solidarity with Victims: Toward a Theology of Social Transformation* (New York: Crossroad, 1982); Lamb, "The Dialectics of Theory and Praxis within Paradigm Analysis," *Lonergan Workshop*, vol. 5 (1985), 71–114, available at: http://www.lonerganresource.com/pdf/journals/Lonergan_Workshop_Vol_5.pdf
[9] William Mathews, "Explanation in the Social Sciences," in *Religion and Culture: Essays in Honor of Bernard Lonergan* (Albany: State University of New York Press, 1987), ed. Timothy Fallon and Philip Riley, 245–260.
[10] Bernard Lonergan, *For a New Political Economy*, ed. Philip McShane, vol. 21, *Collected Works of Bernard Lonergan* (Toronto: University of Toronto Press, 1998); Lonergan, *Macroeconomic Dynamics: An Essay in Circulation Analysis*, ed. Frederick Lawrence, Patrick Byrne, and Charles Hefling, Jr., vol. 15, *Collected Works of Bernard Lonergan* (Toronto: University of Toronto Press, 1999).
[11] Kenneth Melchin, "Exploring the Idea of Private Property: A Small Step Along the Road from Common Sense to Theory," *Journal of Macrodynamic Analysis*, 3 (2003), 287–301, available at: http://journals.library.mun.ca/ojs/index.php/jmda/article/view/131/84.
[12] See generally Melchin, *History, Ethics, and Emergent Probability* and Melchin, "Exploring the Idea of Private Property."

science research sought explanation of events in the motivations and attitudes of social/economic agents, groups or classes.[13]

In the social sciences, much is made of the distinction between fact and value, between descriptive statements and prescriptive statements, and whether ought-statements can be derived from is- statements. A second break-through moment was the discovery that a theory in the social sciences is a theory of some value or other. Value is to be understood as "what is intended in a question for deliberation,"[14] something worthwhile that is intended through a course of action. A value is realized, brought about, created by an individual or a group through sets of activities or sets of sets of activities. This something worthwhile intended includes everything created through human activity such as health and vitality, technologies, economic goods and services, political institutions, common meanings and personal meanings, as well as the structures that facilitate and support them and that constitute an economy, a society, and a culture. It is these values that need to be explained, and they are explained by reference to the set of activities that constitute or bring them about. In other words, this set of activities and their relations is the set of the conditions for the occurrence of this value, this is the set of activities has to occur and to occur in certain relations to constitute or bring about this value. It is normative for the realization of the value.[15]

A third break-through moment was the discovery that a theory answers a what-is-it-question and that a theory of housing is the set of related elements that are relevant, significant, and essential to the constitution of housing. An explanation is an answer to a what-is-it-question. Rather than descriptive definitions, a what-is-it-question heads for and demands an explanatory definition.

A fourth break-through moment was the discovery that a theory of housing is a set of related elements which are variable and admit of a range of possibilities. (This explains why across different countries there are different housing systems.) Further, that what housing is can be distinguished from how it can be used. Housing has some role, purpose, or function within the constitution of other values. These values are constituted by their own sets of related variable elements. The role of

---

[13] See Sean McNelis, "A Prelude to (Lonergan's) Economics," *The Lonergan Review*, vol. 11, no. 1 (2010), 107–120 (2010).

[14] Lonergan, *Method*, 37.

[15] This break-through moment occurred as a result of my seeking to understand the meaning and relevance of a comment by Melchin: "The explanandum of economics is 'value.'" Kenneth Melchin, "Economies, Ethics, and the Structure of Social Living," *Humanomics*, vol. 10, no. 3 (1994), 21–57, 25. It was further provoked by an article by Sauer which addressed seven long-standing dichotomies in social science: fact/value, descriptive/prescriptive, is/ought, positive/normative, ethical neutrality/value permeation, denotive/normative, and cause-effect/means-end. See Sauer, "Economics and Ethics: Foundations for a Transdisciplinary Dialogue," *Humanomics*, vol. 11, no.1/2 (1995), 5–91.

housing within the constitution of these other variables may be direct or indirect. Insofar as housing plays a role in some other value, it has a particular type of hierarchical relationship, a lower value to a higher value. In this relationship, the higher value cannot be achieved unless the lower value is achieved. At the same time, the higher value can order or systematize the particularity of the lower values so the higher value is achieved or better achieved. Through this distinction (between what housing is and its role or purpose in the constitution of other things) we can not only explain what constitutes housing but we can explain an actual operating housing system in terms of (i) a set of related *variable* elements that constitute housing and (ii) a set of other values that order the variable elements in particular ways.[16]

This long 10-year process involved a slow unfolding series of personal discoveries. As a housing researcher enmeshed in a common-sense framework, I found reaching some minimal understanding of each functional specialty a major challenge; I found it very difficult to imagine, to fantasize about something which requires such a fundamental transformation of my thinking and my doing of housing research. It makes demands upon both my self-understanding and upon decisions I make as to whom I will become both as a housing researcher and as person. My challenge throughout this long gestation stemmed from the great difficulty I faced in grasping who I am and what I was doing. I had to come to some understanding of my practices as a housing researcher, then some understanding of what I was doing when I was evaluating them, and finally some understanding of what I was doing when I decided to implement something new.

The context of as-yet functional collaboration

The goal in the functional specialty Policies is to propose a future direction for housing. As a researcher operating in this functional specialty seeking to achieve this goal I am conscious that I am building upon the (as yet to be) work of prior

---

[16] For example, if we consider the set of related elements that constitute a dwelling, it would include elements such as materials and design in certain relations. The particular materials that, in part, constitute a dwelling may, however, vary. They could be one or a combination of timber, stone, bricks, concrete, glass, plastic, steel, aluminum, ice, brush, thatch, snow etc. The particular designs that, in part, constitute a dwelling may include any number of rooms in different configurations with differing purposes according to social, cultural and personal preferences. The range of these materials or designs is not unlimited; rather, there is a range of possibilities such that the particularity of the dwelling may vary.

These last two break-through moments occurred as a result of my seeking to understand "what theory is." Two books by McShane were particularly important: *Randomness, Statistics, and Emergence* (Dublin: Gill & Macmillan, 1970), and *Process: Introducing Themselves to Young (Christian) Minders* (1989), available at: http://www.philipmcshane.org.

This ordering of lower by higher systems mentioned in the text above is my understanding of Philip McShane's 'aggreformic structures' or $W_1, f(p_i; c_j; b_k; z_l; u_m; r_n)$.

functional specialties. To do my work in Policies I need handed-on results from previous specialties.

(i) From *Research*, I need the time-place data as it is ordered by a theory of housing.

(ii) From *Interpretation*, I need a theory of housing that distinguishes those elements and their relations that constitute housing and, that distinguishes and relates a complete range of values in which housing plays some role or purpose (such as a standard of living, wealth accumulation, access to goods and services, status, equity). These roles order the particularities of housing (and the elements and their relations that constitute it). In this way, I can explain the characteristics of any actually operating housing system.

Without this theory I don't have an understanding of what does and does not constitute housing, how each role orders its particularity and how these roles relate to one another in ordering the particularity of housing. Without a theory of housing, I do not have any precise control of my understanding of housing and any policy regarding the future direction of housing will be just guess work; my policy may only be vaguely related to housing; my policy may be asking too much of housing (because it is but one of a number of related things that bring about something different). Indeed, it may even be unrelated to housing.

In my attempt to operate in the functional specialty Policies, I am operating with a very inadequate heuristic of housing; the current operative understanding of housing is implicit and 'the leap' to even an inadequate formulated theory remains to be made. (Once made, this theory will be a developing one.)

(iii) From *History*, I need an understanding of the past and current dynamics (the result of past Policies) that have produced the current actual operating housing system.

(iv) From *Dialectics*, I need an evaluation of these past and current dynamics (the result of past Policies) of housing development, an evaluation that appreciates what these dynamics have achieved yet recognizes how these dynamics have produced variable results—at the extreme very poor housing, slums and homelessness for some people. Here I also need to draw on the experience of the histories of housing in different countries each with their own dynamics of development producing their own results.

(v) From *Foundations*, I need some new appreciation of our sociality, our being-together and an aspiration for the realization of a better future for the whole of humanity.[17]

---

[17] Part II below will illustrate more concretely the relationship between these functional specialties and the functional specialty Policies.

Ib. Prior to Policies—Filling in the 'Functional Specialty' Gaps

The context outlined in Part 1a points to two major problems I face in moving housing forward. First, as a result of my personal context of deep involvement in housing policy and research operating within the practicalities of its taken-for-grant world of everyday common sense, my habit is to propose immediately practical solutions to the problems of housing in Australia (from within my particular viewpoint). So my thinking is in terms of doing this or that practical activity rather than in terms of what value or what future direction will promote the development of housing.

Any attempt at Policies presupposes not just an understanding of the role of this functional specialty within Functional Collaboration but also the results of the five prior functional specialties. My second problem, then, is the 'as yet to be' of the five functional specialties mentioned above.

In this context, I want to propose policies in *two* related areas: first, a new direction for housing research; and second, a new direction for housing. They are related in that future housing directions will depend upon our understanding of housing and that understanding will depend upon the methods we use. The inadequacies of housing are, in part, related to the inadequacies of housing research. Both sets of policies have their basis in Foundations. So, in the absence of the functional specialty Foundations, I will engage in a fantasy in two senses: fantasy as speculation on what might emerge from Foundations; fantasy as envisioning a new future in which Foundations will decidedly formulate a new appreciation of our capacity for performance—our capacity to understand, our capacity to create something worthwhile, our capacity for sociality and collaboration and, our capacity 'to be' in the world.

In trying to forge a new direction in housing research and housing, I have not only to understand and respond to the mess in both but also to work out a trajectory or set of trajectories for their future development. In the absence of Functional Collaboration, I have to somehow 'fill in the gaps' or rather, the chasm. And to do so here, I cannot but operate within an inadequate common sense mode. So, I will begin this Part by discussing and complexifying the analogy of the Dr. House team as an illustration of the 'policy' problem faced in Policies. I will then go on to describe, in the mode of policy analysis often used in housing research, the mess of housing and past policies. This is followed by a section which moves beyond critique to a new appreciation of our capacity to understand and our capacity for sociality as foundations for something new in the future. Finally, on the basis of the foregoing, Part 2 will propose some housing research policies and some housing policies.

A. An Analogy of the Current Mess of Housing and Housing Policies

Recently, Phil McShane used the analogy of the Dr. House team to illustrate the mess with which we are confronted. In the analogy, the Dr. House team is presented with a patient with an unknown illness. A whole range of biological systems are not working, are 'out of sync.' The particular system within which the 'sickness' lies (i.e. whose conditions for recurrence are not met) is difficult to detect because a range of systems is affected and because the failure lies in some part of some system and a whole range of systems depends upon the health of this system. The identification of what is wrong depends upon the heuristic of the Dr. House team, i.e. their current understanding of the human organism as a set of inter-related systems (and sub-systems): circulatory, respiratory, skeletal, nervous, muscular, endocrine, lymphatic, integumentary, digestive, urinary, and reproductive systems and the various sub-systems that constitute each of these systems. Further, the identification depends upon an understanding of the possibilities, the variations and the ranges within which these systems and sub-systems can operate; it depends upon an understanding of all these systems and their inter-relationships, an adequate theory or heuristic of the whole human organism. The aspiration of a healthy organism will be realized when the 'sick' element is identified, when the conditions for the proper functioning of this one element become the policies or future directions for implementation, when these conditions 'fit' with all the systems and when these conditions are actually put in place.

The larger aspiration, however, is for a further development of the human organism. Dr. House's team not only bring an understanding of each single system, an understanding of the sub-systems that constitute each system, an understanding of each single system in relation to all other systems, an understanding of their possible relations, variations and ranges of operation and an understanding of how these systems together constitute the human organism as a whole, they also bring an understanding of these various systems and sub-systems as they change and adapt over time in response to their internal and external environment, as they develop through time as the human organism moves from childhood through to adulthood and an understanding of the shifting dynamic of development.[18]

It is in this context that they propose and implement a solution (or when one solution is not effective to go and propose another solution) to restore the sick human organism health and on the path to further development. That solution will depend upon the stage of development of the human organism.

Understanding the mess of housing and past policies

One of the symptoms of the current mess in housing, even in countries with developed economies, is the difficulty that a significant proportion of households (individuals, groups, families) have in finding housing that is affordable, in good condition and appropriate to their needs: at the extreme, some households are

---

[18] It is a shifting dynamic of development, as I would suspect that at different 'stages' of development, one or the other of these various systems 'takes a leading role.'

homeless; others live in slums, in sub-standard, unhealthy and even dangerous housing; some live in housing poverty struggling to pay their rent or mortgage repayments; some live in housing without adequate water supply and sewerage, without adequate facilities, without ventilation, heating and cooling; some live in housing located in places that have poor physical and social infrastructure, that have poor access to employment, to medical, educational, retail and recreational services; some are asset-rich living in grand houses but income-poor and so unable to maintain their housing etc.

One pervading context for this mess in housing seems to be the dominant dynamic[19] of wealth accumulation or profit maximization. Throughout the housing system—the ownership/control, rezoning and development of land, the design of neighborhoods and housing, the construction, rehabilitation, reconfiguration and conversion of dwellings and, the exchange of land and dwellings—wealth accumulation is the dominant motivation and housing is tailored according to the extent to which different groups with different wealth/income can serve this end. Within this dynamic, financial viability is not an adequate standard. Rather, the standard is higher—selective investment in those enterprises in which, it is anticipated, returns will be maximized.

Thus, the dynamic of 'wealth accumulation' increasingly compromises different aspects of housing, such as: the ownership and use of land; the impact of building materials, housing design and housing utilities on the environment; housing standards and quality; the establishment and maintenance of neighborhood communities; the social demands of higher density living; security of tenure; the exclusion of people from decisions; and, personal space 'to dwell' etc. But housing is not a simple entity. It incorporates a range of elements. While 'wealth accumulation' forms the context for change in housing, change also occurs in each of the elements.

At different times in the history of housing, some aspects of housing have been more predominant than others while other aspects have receded. Various dynamics throughout the 19th and 20th century in developed economies have brought about change in housing. In the 19th century, poor housing standards and the proximity of dwellings to noxious and malodourous industries threatened the health not just of local inhabitants but also the wider community. A strong planning movement led to the introduction of land use planning legislation (separating areas for residential land uses from other land uses) and minimum building standards legislation. Throughout the 20th century, the increasing efficiency in the housing construction industry led to new building techniques, new building materials, new building types (such as high rise living) and new building technologies. In the latter half of the 20th century, the desire of households to purchase their housing (rather than renting)

---

[19] As noted above, a dynamic is the result of past Policies.

spawned a series of financial innovations and, the failure of the housing system to provide housing to significant portions of the population produced a social housing movement. An environmental movement has sought to preserve, respect and enhance our environment by considering the impact of housing on the natural and ecological environment and changing the materials used in housing construction, the siting, design and environment of dwellings and the types of supporting infrastructure (water, electricity, gas, sewage etc.) etc. Self-help and participatory democracy movements continue to promote housing co-operatives, neighborhood groups, tenant groups, self-management groups and various forms of community organizations advocating for better housing conditions. Different religious beliefs find expression in particular aspects of the design, form, and use of housing. For example, for some people, concern and action around housing poverty is an expression of their solidarity in the Kingdom of God; feng shui is often used to determine the design and layout of dwellings; some religious beliefs find expression in the separation of spaces for men and for women, in dietary requirements, in separate spaces for food preparation and in the inclusion of a prayer room, a shrine, religious hangings, and adornments within a dwelling.

This short list of dynamics highlights the various dimensions of housing: environmental, technological, economic, political, cultural, and religious. Some aspects of housing have changed as a result of these dynamics. Around the dominant dynamic of 'wealth accumulation' and each of the secondary dynamics (and there are many more of them) there coalesces a movement which accentuates the contribution, the achievement and the possibilities of each and downplays the negatives. In this narrow interest lies the bias of the group.

One housing issue of particular concern to housing researchers is housing poverty. Various policies have been adopted to address this issue: shelters for the homeless which may also include a range of services to meet other health and social needs; public housing; housing provided by community organizations, by charities, by foundations and trusts etc.; government income support (including rental/housing assistance); incentives to direct private sector housing to low-income households such as tax credits; planning and building regulation etc.

Indeed, what becomes apparent as we consider the pre-suppositions of these various policies is the extent to which these policies reflect an understanding of the 'causes' of housing poverty. A housing policy may attribute poor housing to the inadequacies of individuals—they are lazy and are not prepared to work; they are sick, have a terminal or debilitating illness, have a disability (whether a physical, intellectual or sensory disability) or, have a mental illness; they belong to certain ethnic or cultural group; they are too fussy and expect too much, etc. Or, the housing policy might attribute housing poverty to the inadequacies of an economic system that discriminates against certain households. The latter housing policies can range from those that 'complement' the current economic system (with some compromises) through those that seek to shift the grounds on which the current economic system is based (such as through housing co-operatives) to those that seek the abolition of the current economic system (such as through state confiscation of private property).

To varying extents these policies (presupposing different understanding of the 'causes' of housing poverty) have been tried over the past century and have become entrenched within different groups and classes as the best way forward. We are then confronted with the question: what is their contribution to the development of housing, housing-as-a-whole? To what extent have the interest of particular groups and classes over-estimated their contribution and distorted the development of housing?

Indeed, reflecting on questions such as these, we can discover that these responses to housing poverty (i.e., past housing policies) not only reflect the 'causes' of housing poverty, they are also an expression of an understanding of humanity, an understanding of myself/ourselves/others, an understanding of our relationships, a vision for the future, for what we can become etc. They not only reflect an acceptance of households living in housing poverty, they reflect the attempts of others to maintain their current way of life—living in luxury and accumulating wealth—and attempts to promote this as good. As such they can become a form of personal aggrandizement; housing becomes a status symbol; exclusive suburbs or gated communities promote the formation of groups with a common culture and their separation from other groups (whose different culture can disturb the equilibrium of the group). While a few households with wealth and income can realize these directions, most just aspire to them and in doing so these directions have become taken-for- granted worthwhile housing directions.

Like the Dr. House analogy, the disease of housing lies within one sub-system among a number of sub-sub-systems, sub-systems and systems that constitute an economy, a society, a culture, a personal identity. Housing poverty is a symptom of many sub-systems 'out of sync.' The question remains as to how to go forward. What is the intelligent, reasonable, responsible and adventurous thing to do that will not only restore the health of housing, of the economy, society, culture and personal identity but will also take the next step in its development.

B. Implementing a New Appreciation of our Capacity for Performance: Foundations

While Dialectic discovers something new about our capacity for performance, Foundations is grounded in the decision to operate on this basis. The following seeks to treat this decision as fantasy about the future. Here, the fantasy results in deciding to implement a new appreciation of our capacity for performance in understanding, in collaboration and in dwelling in the world.

1. Deciding to implement a new appreciation of our capacity for understanding

Policies to be proposed in Part 2 have their grounds in a decision to implement a new appreciation of our capacity for performance in four aspects of understanding:[20]

  i. the (re)discovery of theory or explanatory definition;

  ii. an integration of the disciplines;

  iii. an integration of the diverse methods used in housing research; and

  iv. a new heuristic of the economy.

First, the question, what is it, is answered not through describing the characteristics or purpose of something but by explaining the occurrence of something in terms of the conditions necessary for its occurrence, i.e. the set of elements and their relations that constitute its occurrence.[21] It moves beyond the associations in time and place that dominate current social sciences to systematic correlations.[22] It results in a more precise grasp and control of what is under consideration.

One of the tasks of the specialty Doctrines is to make precise the content of popular doctrines: the precising is quite remote, as are the results of all the specialties, from commonsense meaning.[23]

This new appreciation of our capacity to understand poses some difficulties for commonsense understanding. The dynamics of housing development (see above) are values that different groups seek to realize, they are "what is intended in a question for deliberation"[24]—will I realize this value?

Within the world of common sense, these dynamics are real. In the world of theory, however, their reality remains something yet to be determined. In this world, we raise a question about the value we are seeking to realize. Is it? Is it real? What constitutes this value? What is intended by deliberation? For instance, while commonsense understanding talks about 'wealth accumulation', in the world of

---

[20] In *Insight*, Lonergan takes us on a personal journey of discovering our capacity for understanding in increasingly complex ways. Beginning with some simple illustrations, he asks us sequentially to distinguish theoretical and commonsense understanding, genetic understanding, and dialectical understanding.

[21] We thus define something in terms of the *method* of its constitution.

[22] This type of thinking, which seeks to associate events in time and place, dominates both quantitative and qualitative research in the social sciences. This, I think, is one of the points that Terry Quinn is making in his chapter on "Interpreting Lonergan's Fifth Chapter of *Insight.*"

[23] Philip McShane, *Pastkeynes Pastmodern Economics: A Fresh Pragmatism* (Axial Publishing, 2002), 78.

[24] Lonergan, *Method in Theology*, 37.

theory we are seeking for the elements that constitute wealth accumulation. Without such a theory, I cannot then go on to consider whether 'wealth accumulation' is, i.e. whether it is real. It certainly is within a common sense framework but the real issue is whether (and in what way) it is a value.

The difficulties inherent in such questions are illustrated by considering the many households at the extreme of housing poverty, those who are homeless. Homelessness describes a situation in which many households find themselves. It is a descriptive understanding not an explanatory understanding which would identify the set of elements and relations that constitute homelessness. However, what becomes evident as we seek the intelligibility of homelessness by seeking an answer to the question, what is homelessness, is that homelessness is not intelligible. It is part of the social surd "that (1) is immanent in the social facts, (2) is not intelligible, yet (3) cannot be abstracted from if one is to consider the facts as in fact they are."[25] Thus, we cannot arrive at an explanatory understanding of homelessness as such, for it is not real, it is not something that is intended by deliberation (and we can only intend for deliberation some value or other—we cannot intend evil!). Rather homelessness is to be understood as the absence of something else, the absence of some value that is intended by deliberation. We can only arrive at an understanding of homelessness by understanding something else. For Policies, the import of this is that homelessness is not eliminated directly but through the promotion of something else, something that is worthwhile.

Thus, the functional specialty Policies relies upon explanatory understanding which is reached in the functional specialty Interpretation, as it (Policies) intends value rather than some vague common sense view of what is wrong with the current situation.

Second, there is the question as to how the various dynamics (referred to above) are related to one another and to housing? Indeed, what is the broader context within which housing plays a role—in an economy, in a polis, in a culture, in the ultimate context of the Kingdom of God? So, for instance, in the example of housing poverty (referred to above and below), current responses range imprecisely across these different dimensions. By relating these various dynamics together within the broader context within which housing plays a role, we can more precisely locate where the problem lies: some aspects of the problem are economic (inadequate responses to phases in the trade cycle) whereas as others lie in the polis (the decisions whereby the costs and benefits of economic production are inequitably distributed within the population), in a culture (owner-occupation as the preferred tenure) or in a religion (it is what it is because God ordains it so).

It is the role of the functional specialty Interpretation to specify the range of possible relations between housing and its broader context. It is this specification

---

[25] *Insight*, *CWL* 3, chapter 7, Section 8.2, 255.

that will account for the differences in the characteristics of an operating housing system between one country and another.

Third, while housing research has utilized a broad range of methods, their integration has been problematic. Functional Collaboration draws these methods together in a new way, refocusing them as the ways in which answers can be found to a complete set of eight questions.

Fourth, Lonergan's writing on economics present a new heuristic of economics. The critical point of difference here is that this heuristic is based upon a quite precise understanding of what it is to understand. This new heuristic is an explanatory definition or theory of economics in a context which rigorously distinguishes economics from technology, politics, cultural studies etc.[26]

2. Deciding to implement a new appreciation of our capacity for solidarity, collaboration and dwelling in the world

Part II on reaching for Policies also has its grounds in a decision to implement a new appreciation for our capacity for performance in collaboration and in dwelling in the world. As Michael Shute in some recent articles has indicated, collaboration has a long history.[27] However, here a new appreciation of collaboration (which has its roots in our sociality) is emerging in the face of centuries of individualism. A new appreciation of our capacity for 'dwelling in the world' is reflected in Aaron Mundine's chapter on "Functional History and Functional Historians" with its cultivation of a relaxed pace.[28]

---

[26] It is this aspect of Lonergan's economics that many economists such as Paul Oslington have failed to grasp. For example, see Oslington, "Lonergan's Reception among Economists: Tale of a Dead Fish and an Agenda for Future Work," *METHOD: Journal of Lonergan Studies*, n.s., 2(1) (2011), 67–78, and see as well his article (with Neil Ormerod and Robin Koning), "The Development of Catholic Social Teaching on Economics: Bernard Lonergan and Benedict XVI," *Theological Studies* (2012), vol. 73, no. 2, 391–421. The same aspect also explains why Michael Shute has experienced great difficulty communicating Lonergan's economics. See his chapter below, "Communicating Macroeconomic Dynamics Functionally."

[27] Michael Shute, "Functional Collaboration as the Implementation of 'Lonergan's Method': Part 1: For What Problem Is Functional Collaboration the Solution?" *Divyadaan: Journal of Philosophy & Education* (2013), vol. 24, no. 1, 1–34, and "Functional Collaboration as the Implementation of Lonergan's Method: Part 2: How Might We Implement Functional Collaboration?" *Divyadaan: Journal of Philosophy & Education*, vol. 24, no. 2, 159–190.

[28] This harkens back to a traditional notion of complacency. See Frederick E. Crowe, "Complacency and Concern in the Thought of St. Thomas," *Theological Studies* 20 (1959), 1–39, 198–230, 343–395, republished in Crowe, *Three Thomist Studies*, Supplementary Issue of *Lonergan Workshop*, vol. 16, ed. Frederick Lawrence (Boston: Boston College Press, 2000), 71–204.

II. Reaching for Policies: New Directions in Housing Research and Housing

The results of the functional specialty Foundations are the thematization of the horizon within which future decisions will implement a new appreciation of:

i. our personal and collective capacity for performance in understanding, one which distinguishes and integrates a range of different questions and the respective methods by which these questions are answered; and

ii. our need to dwell in the world.

This fantasy about creating something new in history is passed on to the functional specialty Policies.

The functional specialty Policies proposes that this fantasy can be implemented through the selection of certain values.

The housing researcher operating in the functional specialty Policies faces three problems:

i. in moving forward, a housing policy will, in some way, incorporate a series of discoveries—of what is and is not relevant to housing;

ii. if we are to chart a way forward for housing by proposing a new direction for its development then we have to come to terms with each of the questions posed for Research, Interpretation, History, Dialectics and Foundations, and;

iii. it is only within the context of answers to these questions that the researcher operating within the functional specialties can (i) grasp the current dynamic of development of housing (in its larger economic, political, cultural and religious contexts) and, (ii) make the adventurous leap (which is also attentive, intelligent, reasonable and responsible) into proposing/prioritizing/re-prioritizing the value or dynamic which will promote the further development of housing.[29]

A. Housing Research: A New Methodological Framework

Operating in the functional specialty Policies I propose that housing research operate within the framework of Functional Collaboration which distinguishes a

---

[29] So now, I hope, you can appreciate my problem as I seek to operate in the functional specialty Policies. I do not have an adequate heuristic of housing, i.e., an understanding of housing as a set of inter-related systems (and sub-systems) and its possible contribution to (or relationship to) an economy, a society, a culture, and the Kingdom of God. This is the function of Interpretation. Indeed, I don't have a solid grasp of the dynamics of an already operating housing system. This is the function of History. I don't have a solid grasp of the contribution of these various dynamics to the progress in housing, etc.

complete set of eight inter-related questions and the methods by which each of these questions is answered.[30]

Further, I propose that we will reach a more adequate understanding of housing by developing an explanatory definition or heuristic in which is grasped (i) the significant, relevant and essential elements and their relations that constitute the occurrence of housing, and (ii) the ranges of possible relationships between the particular characteristics of housing and the function that housing plays in the achievement of other values (such as various economic, political, cultural, personal and religious meanings).[31]

In addition, I propose that a more adequate understanding of the economics of housing will distinguish between, on the one hand, the production of new housing within the two circuits of production (basic and surplus) with its final sale contributing to the standard of living and, on the other hand, the exchange of existing housing.[32]

Finally, I propose that the functional specialty Policies asks the question: what new operative dynamic or vector will best promote the future development of the current actual operating housing system into future systems? The functional specialty Policies will focus on the housing system as whole, on the totality of its purposes.[33]

---

[30] Explanatory note: this method (i) relates and restructures a range of disparate methods that are currently haphazardly used in the social sciences, and (ii) distinguishes specialties within the process of moving from the current situation to implementing something new while maintaining the whole of housing throughout the process.

[31] Explanatory note: the current mode of understanding operates within the horizon of common sense where the occurrence of housing is understood in terms of the motivations and interests of the major actors—planners, land developers, builders, estate agents, governments, consumers, and other political interests. See, for example, the work of Ball and Harloe as outlined in chapter four of my 2014 book, *Making Progress in Housing*.

[32] Explanatory note: the current mode of understanding does not distinguish between the construction and sale of new housing and the sale of existing housing. See Lonergan's theory of the economy in *For a New Political Economy*. See also McShane, *Economics for Everyone: Das Jus Kapital* (Halifax: Axial Publishing, 1998), chapter 3, and McShane, *Futurology Express* (Vancouver: Axial Publishing, 2013).

[33] Explanatory note: Current discussion of housing policy operates from a particular viewpoint, whether governments or organizations, and what they want or what others want from housing. It is policy in the interests of a particular organization or group. In contrast, the functional specialty Policies uses a more adequate heuristic of housing developed by the functional specialty Interpretation (see above), grasps the vectors within the current actual operating housing system as understood by the functional specialty History, and draws on the appreciation/critique of the functional specialty Dialectic to propose new directions or vectors that bring about future development of housing as whole. It is a vector through

B. Housing Policy

I propose that the vector of 'dwelling in the world' will take housing to its next stage of development. Housing plays a role in constituting this vector and, as such, the characteristics of housing will need to be adapted to achieve this vector.[34]

Descriptively I would understand that housing would contribute to 'dwelling in the world' by providing a place of 'privacy and security,' where there is 'the possibility of intimacy,' where 'intimacy can be protected,' where 'we can be complacent—take what we are and what we have for granted—where we can 'exercise some control' and make choices, where we can express our self and where we share 'a common interest based on the shared experience of housing itself.'[35]

III. The Functional Specialty Polices Handing on to the Functional Specialty Systematics

How can these methodological Policies for housing research contribute to a better understanding of housing? How can housing contribute to the constitution of 'dwelling in the world' given the current ranges of technological, economic, political, cultural and religious requires many adaptations in the characteristics of housing.

Here I am passing on some policies for the future development of housing research and housing to Systematics. The question for Systematics is: what course of action will integrate these new vectors or policies within the complex series of contexts that constitute housing research and an actual operative housing system at the present time? The answer to this question will integrate these new vectors within these ecological, technological, economic, political, cultural and religious contexts.

So, for instance, Systematics would consider the vector of 'dwelling in the world':

i.   in the context of climate, the land on which it is located, the materials extracted etc. already taken-for-granted by land and housing developers, natural resource industries etc.;

---

which not only housing will be better realized but also the full range of purposes in which housing plays a role.

[34] Explanatory note: As noted previously, a range of vectors are currently operative within housing, some more dominant than others: wealth accumulation; land-use planning; enhancing the natural and ecological environment through changes in materials, technologies, site design, physical infrastructure; financial mechanisms and instruments; self-help and participatory democracy, etc. By proposing a vector of 'dwelling in the world,' I am proposing that the particular characteristics of these vectors are adapted in such a way that this value is achieved.

[35] See Peter King, *In Dwelling: Implacability, Exclusion, and Acceptance* (Hampshire, England: Ashgate Publishing, 2008).

ii.   in the context of the materials used for building, tools, design, siting etc. already taken-for- granted by producers of building materials, builders, planners and architects;

iii. in the context of how housing is produced and exchanged;

iv.   in the context of how housing is distributed and the conditions under which they are occupied;

v.   in the context of the cultural meanings of relationship with the land, location, architecture, etc., and

vi.   in the context of the religious meanings of 'dwelling in the world.'

IV. Some Concluding Personal Reflections on What I Have Done

If there is any movement here in doing Policies, it is in identifying the problem a bit more clearly. Indeed I find myself somewhat overwhelmed by the task given the absence of at least a theory of housing. So, as you see, I still have some work to do on the guts of Policies. There, in Part II, I've only indicated in a few paragraphs some future directions for housing research and housing stemming from Foundations, a decision to implement a fantasy of a new appreciation of our capacity for performance in human understanding and in human solidarity, collaboration and dwelling in the world.

Throughout Part II, the ongoing struggle was to identify what it means to operate within the functional specialty Policies, in particular to distinguish this from practical proposals that would emerge in the functional specialty Communications. Moreover, it is a struggle to identify the method of Policies, and it leads me back to further work on what is happening when making judgments of value[36] as bringing about new possibilities. However, the surprising feature, at least to me, was the extent to which the functional specialty Policies drew not only upon the results of the functional specialty Foundations but also drew upon the results of other functional specialties: Interpretation for a heuristic of housing; History for an understanding of the current policies/doctrines already operative within an actual housing system; Dialectic for an appreciation and critique of these current policies/doctrines. The results of these functional specialties provided the 'poise' within which to project a way forward for the development of housing (as a whole).

Finally, I would note the speculative nature of what I have done… it stretches beyond where we are now into the land of new possibilities… a land of beyond imagination… a land of Policies guiding the future of the globe.

---

[36] See chapter 2, section 4, of *Method in Theology*.

William Zanardi

The conference that led to this volume had a purpose: the reorientation of Lonergan studies. The basic question was how to achieve this, and the various papers are an assembling of related and incomplete answers. My task is to focus on strategies of implementation that belong specifically to the eighth functional specialty. What might they be? Presumably identifying them is part of the "performance" of Communications, but, if I lack the needed insights and familiarity with their "range of relevance," what is to be done?[2] Fred Crowe offered a realistic answer: "When you have a mountain to move, and only a spade and a wheelbarrow to work with, you can either sit on your hands or you can put spade to earth and move the first sod."[3] The modest recommendations of this essay are a bit of initial sod turning.

Following the common format of the preceding contributions, I begin with some introductory remarks on my current understanding of my task and why I think it's worth the effort. Next is a section specifying "content" that may be strategically transformative of Lonergan studies. A third part comments on how the content might be handed on to two, even to three, different audiences. Finally, I offer some remarks about the credibility of our early ventures in doing functional specialization.

I. My Current Appraisals

What is my current understanding of the meaning of Communications? It appears to be a Janus-faced enterprise of communicating, first, what is received from prior specialties to audiences unfamiliar with functional specialization (hereafter FS). McShane has been assigning this task to C9; yet currently this is what we are all doing in the absence of sufficient practice with, an adequately developed understanding of, and a substantial body of work for modeling FS. Second, FS8 will someday be a science of communication that fills a specialized role in collaborative inquiries, primarily by applying policies and plans generated in the preceding specialties and

---

[1] "Identification is performance. Its effect is to make one possess the insight as one's own, to be assured in one's use of it, to be familiar with the range of its relevance." *Insight, CWL* 3, 582.
[2] The question is a nod to Lenin's predicament when he had to muddle through a post-revolutionary period without guidelines from Marx. A less controversial parallel: the stranded crew in a lifeboat on a cloudy night might ask, "Which way is the nearest shore?" Without an answer, their efforts in rowing may be aimless.
[3] *Theology of the Christian Word: A Study in History* (New York: Paulist Press, 1978).

by tracking and reporting results to colleagues within FS8 and to the other specialties.[4]

Why make an effort to speed up the emergence of this science? Well, how important is it to make more effective local and global plans for achieving good ends?[5] If the purpose of such a science is to spread promising ideas across diverse populations and to track and report on the results of their implementation, might its emergence contribute to the improvement of human history?[6] Pat Brown's paper on exponential growth is a slice of a larger perspective on the promise of FS for the control and improvement of history. Is it no more than a dream? If the division of labor already exists in some enterprises and if "what might be" is part of the whatness of a thing, then this unrealized possibility is not unrealistic.[7]

The goals are high, but the means at hand are paltry. FS8 seeks in its own ways "to solve problems, to erect syntheses, to embrace the universe in a single view."[8] However, its ways of improving history are inchoate, a tadpole's dreams of becoming a mature frog adding to the beauty of a nightly chorus. So my appraisal is that Communications as a science is worth pursuing even if, for now, our efforts are confined to C9, chatting among ourselves but with an eye to more serious exchanges someday.

---

[4] Why only "someday" a science? Communications is plausibly the hardest of the eight specialties. Among undergraduates it is a far more popular major than physics. If we were all undergraduate majors in the former, we would not need to understand anything about FS, our roles as individual contributors and what our relation as a community of specialists is to the meaning of history. As well, the "theories" of communication surveyed in our textbooks would be noticeably pre-theoretical. (Would anyone really notice?) Many of the differences among the theories would reflect implicit differences among philosophical views of reality, knowing and objectivity. (Would anyone be excavating those hidden origins?)

[5] "In so far as it is effective [Communications] persuades or commands others or it directs man's control over nature" and history. *Method in Theology*, 356.

[6] But what do we mean by "improvement"? McShane has posed in various places a basic and cleverly worded challenge to those fumbling for an answer. "Do you view yourself and humanity as possibly maturing—in some serious way—or just messing along between good and evil, whatever you think they are?" Adding "yourself" to the question is both clever (Who wants to say "no"?) and subtle. To self-attend and discover hidden potential may be to move toward generalizing from oneself to the "obediential potency" of the entire species.

[7] McShane *Question* 56, "Breaking Forward to Global Care," footnote 14. This essay is available at: http://www.philipmcshane.org/questions-and-answers.

[8] *Insight, CWL* 3, 442.

II. Content of This Communication

What's to be communicated? Put simply: "This Lonerganist movement has been ineffective, so isn't it time we try something different?" Is the negative judgment too harsh? Given the posthumously uncovered, life-long interests of Lonergan in history and economics, we can reread his remarks on Cosmopolis and his discovery of FS as searchings for a new good of order. Has any sign of that order appeared fifty some years on since his 1965 discovery? Has Lonerganism made any significant differences in the "set ups" of the academy, the economy or any institution? Assuming the current set ups as "flows of goods" are far less than they might be, "ineffective" is an appropriate label for any movement that espouses to make a difference in them but has not after nearly half a century.[9]

What might we try differently? There are clues. "Doing method fundamentally is distinguishing different tasks, and thereby eliminating totalitarian ambitions."[10] Presumably in attending this conference, we are all concerned about the neglect of Lonergan's 1965 discovery because we see in it a structured approach to making care for global meaning more effective. Our task, then, is to persuade others, especially new students of Lonergan, to end the neglect and to exploit the possibilities of FS.

How do we persuade others this division of labor is worth a try? The dynamics of persuasion include arguments but only in a secondary role. The primary role is filled by the affect-laden symbols and signs that inspire a vision of what might be and so attract responses.[11] This primary role can be filled by figures of incarnate meaning who have "been to the mountain" and seen what might yet be. But even they will face difficult challenges. In this limited space, I'll focus on just two of them.

If many academics are career-oriented in set ups that demand rapid productivity, conventionally measured by so many papers and books per time interval, how do we effectively change their minds and hearts? The needed changes (e.g. accepting the slow pace in acquiring serious understanding of anything) amount to a displacement from familiar routines. But who wants to be a displaced person, to be temporarily homeless?[12] Might that fear be what keeps some academics from

---

[9] Lonerganism has not implemented Chapters 14 of both major texts and has not taken seriously 17.3 in *Insight*. Diagnosing why not, and also why forty plus years of efforts have had so little impact, is a task, I suppose, for the historians and dialecticians.

[10] "An Interview with Fr. Bernard Lonergan, S.J." Philip McShane (ed.) in *A Second Collection* (Philadelphia: The Westminster Press, 1974), 212.

[11] "… the basic problem is to discover the dynamic images that both correspond to intellectual contents, orientations, and determinations, yet also possess in the sensitive field the power to issue forth not only into words but also into deeds." *Insight, CWL* 3, 585.

[12] The homeless metaphor is not without precedent. Entry into the horizon of theory in a serious way is time consuming and short of immediate gratification. Most of us in the Humanities can annually write descriptive papers and find homes for them in journals that would reject more theoretical pieces that might "be over the heads of their readership."

asking themselves further questions about history and any roles they might play in improving it? Or even more narrowly, is it threatening to ask if the good of order I am part of is really all that good? So the challenges here are posed both by entrenched practices and by psychic defensiveness. How might we meet both? McShane has offered some advice that, while I don't fully understand it, is likely to be on target.

In numerous places he has written about the importance of "linguistic feedback" in pushing ourselves beyond immature states of self-attention.[13] What is it? I don't claim to have more than a few provisional guesses. He frequently cites what Lonergan writes: "At a higher level of linguistic development, the possibility of insight is achieved by linguistic feed-back, by expressing the subjective experience in words and as subjective."[14] (That sounds like a recipe for exclusion from much of conventional academic discourse.) Still, some writers report that they write their way into insights, so a feedback metaphor seems plausible. However, this reading of what he means won't allow us to escape predictable criticisms (and personal resistance). Why not? McShane says linguistic feedback is to be part of public communication that manages to reach the *moi intime* of both self as writer and the audience as readers. This proposal definitely will evoke professional censure. So why does he keep pushing the proposal? My best guess to date has to do with what appeared above: (1) the dynamics of persuasion include arguments but only in a secondary role; (2) the primary role is filled by the affect-laden symbol that inspires a vision of what might be and so attracts responses; (3) how do we effectively change minds and hearts? When the way forward requires displacements, in some cases painful departures from psychically comfortable surroundings, affective symbols and signs must supplement any arguments if persons are to move toward the exits. So linguistic feedback (*cor ad cor loquitur*) as a personal recounting of one's own doubts, missteps, breakthroughs and even breakdowns may be an effective way of reaching others and inviting them to depart.[15]

That is a beginning in persuading others, but the next question is what effective structure could sustain ongoing efforts to promote FS. We need to "replace casual

---

[13] I suspect this might be a neglected part of what it means to be at home in the third horizon. I can recite the distinctions among intentional acts, their relations among themselves and their proper objects. However, is there a further step in self-luminosity I've been missing? Is a first step exactly what Lonergan meant by self-appropriation when he invited readers to work through relatively simple puzzles? Is a second step, "stabilizing" such experiences and expanding our understanding of them, taken by expressing these experiences "in words and as personal"?

[14] *Method in Theology*, 88 at note 34.

[15] Concrete examples of this sort of self-revelatory communication occur in drug rehabilitation "sharing sessions," the Confessional Literature of the West and the autobiographies of some scientists.

efforts to do good … with a structured grouping of executive reflectors and their operations at the fringe of metaphysics."[16] McShane's analogy is that, while many hands go into the making of cars, it is marketing and sales that make the earlier labor pragmatically purposeful and ongoing.

So how do some of us become effective in marketing and sales? The practical task is inventing or adapting effective strategies. The challenges are positive and negative. How are advances in thinking and practice (new products) to be pushed forward (in expanding markets) and their opposites (obsolete products) left behind? In regard to the second half of the formulation: How do we avoid being "endlessly in pseudo-dialogue with amateurs, however gifted, in the past"?[17] How do we escape the "stale academic ethos" that rewards conventional forms of inertia but, in the Humanities at least, yields no "cumulative and progressive results"?

The answers at first must be fairly general.[18] It helps if salespeople have used their products, believe in them and want others to benefit from similar usage. It also helps if sales and marketing personnel cooperate in promoting these public goods. In contrast to competitive workplaces that pit sales staff against one another to maximize their efforts, imagine a good of order that promoted personal relations.[19] I'll return to this fantasy in part four. For now the question of "content" is about effective sales pitches, about strategies for moving products off the academic shelf and into public markets.

What efforts are likely to be effective at the beginning of a new science? More specifically, what are some promising, concrete strategies for implementing FS well before it is accepted practice?[20] Do we know any pedagogical tricks for infecting the next generation of Lonergan students with doubts about current practices and for arousing their interest in trying something new? Are any teachers among us in a position to incorporate FS into existing courses or even to offer a course on one of

---

[16] McShane email January 5, 2014.

[17] McShane email July 7, 2013.

[18] If Communications is largely a matter of selecting strategies for effectively passing on the results of prior specialties, can they be all that concrete at present if we're "really writing towards the future" (email McShane September 24, 2013) though in continuity with the "genetics of mind seeded by Aristotle"?

[19] "It is useful to have a theoretical structure, to be able to speak of the good of order generally, but the simplest and most effective apprehension of the good of order is in the apprehension of personal relations." *Topics in Education* (Toronto: University of Toronto Press, 1993), *CWL* 10, 41. An article I published in 2010 was inspired in part by reflections on the third line of the diagram on page 48 of *Method in Theology*. That article began with a remembrance of my grandparent's restaurant and the familiar employees who worked there for many years. Why did they stay so long?

[20] Below I offer some randomly accumulated answers to the following series of questions. One hope for the conference is that more novel strategies may appear along with some practical steps toward collaborative experiments.

the specialties? Are any of us up to studying with close attention disciplines in which collaboration and the division of labor are advanced or at least emerging (thereby supplying an inductive argument[21] for FS)? Can we show how the organization of FS is present implicitly in our own performance when we are sufficiently attentive to it?[22] It is possible to teach by negative examples. I once reviewed a special issue of the American journal *History and Theory* devoted to environmental histories, a review that identified and criticized the unintentional mixing of multiple specialties in a single article with very uneven results. This strategy has its risks.[23]

To conclude this section, let me voice one practical stance. If "identification is performance," then Communications is primarily a matter of first inventing strategies for effecting change, then experimenting with them and finally having a communal forum for sharing and assessing the results. The proof of the effectiveness of any strategies will lie in the doing. Experiments with any of the eight specialties (as opposed to the preliminary phase of talking about what they might be) need to be tried, even if found wanting, as opposed to being thought too difficult and so not tried.[24]

III. Effective Strategies for Handing On

First, who are the audiences I have in mind? Ideally, as with dialecticians, there will someday be others in FS8 who will receive and criticize my "content." Recycling at this stage will refine what is recommended as courses of action. Whatever plans are acted upon will need monitoring with the results reported to the broader audience

---

[21] "The deep failure of views and efforts in induction is the failure to self-discover through Lonergan; the deep failure of Lonerganism is a failure of real scientific induction." McShane email July 31, 2013.

[22] For example, we can repeat Lonergan's definition of metaphysics, but might it help to parallel "conception" and "affirmation" with acts of understanding and judging and then to point out that "implementation" parallels acts of deliberating and deciding? The next step pedagogically would be to take commonplace examples of problem solving and to track the range of means and ends in each and how a "division of labor" is not all that uncommon. McShane's example of the family planning a vacation under changed circumstances from previous years will be familiar to most of my readers.

[23] The journal editor responded with lengthy comments but was unwilling to publish the review that, in hindsight, I recognize was not a model of diplomacy.

[24] Last year I published a book, *Comparative Interpretation: A Primer* (Austin: Forty Acres Press, 2013) as an introduction to Lonergan's Dialectic. The target audience was a graduate student population. This year a second book, *Further Case Studies in Comparative Interpretation* will appear, possibly co-authored with one or two graduate students interested in doing FS4. The practical focus of my efforts in both is to reach a new generation of Lonergan students before all sorts of mortmain set in.

of specialists beyond FS8.[25] A third audience is envisioned as those to be persuaded to act on the recommended plans.[26]

Presumably different audiences will receive according to their capacity to receive (*Quodquod recipitur ad modum recipientis recipitur*). Fantasize that members of the first two audiences share a common frame of reference, a standard model for what they are doing. Besides a grasp of their distinct but related roles in the larger enterprise of FS, they have some understanding of the genetics of human meaning regarding their shared topic and so are able to detect if results contribute to progress or to decline.[27] Further fantasize that, when all the operators in the specialties are such personal "contexts," recycling will be far less burdened by having to debate "obsolete products" or to engage in endless "pseudo-dialogues with amateurs."

What of the third audience? Not surprisingly its members will be unfamiliar with Lonergan's work. Among most of them the reigning imperative "Be practical!" will discourage any lengthy ventures into the second and third horizons. Invitations to do so would seem too strange and unrealistic.[28] So the outreach of C9 is to a common-sense orientation, and persuasion will require a "sales staff" adept at making policies and plans plausible and attractive. How can this be done? The needed skills may already be in hand. Some specialists in FS8 will already be well practiced in reaching undergraduates or have backgrounds in pastoral theology or show a knack for publishing in popular media or have some experience with an area of applied ethics and its pre-theoretical discussions of cases. Some may even have had administrative experience in faculty governance and know how to "sell"

---

[25] A linear conception of FS has the "data" accumulated by monitoring reported back to FS1. However, the circulation of results will be far less neat since promising outcomes may affect works-in-progress among the other specialties.

[26] "There is the far more arduous task (1) of effecting an advance in scientific knowledge, (2) of persuading eminent and influential people to consider the advance both thoroughly and fairly, and (3) of having them convince practical policy makers and planners both that the advance exists and that it implies such and such revisions of current policies and planning with such and such effects." *Method in Theology*, 366–367.

[27] In other words, they are familiar with the universal viewpoint. While this phrase, in an elementary sense, usually means an understanding of the critical realism associated with questions of judgment, an expanded meaning includes the demands of theory for ever more comprehensive understanding and a situating of such demands within an adequate philosophy, even a theology, of history. Part 4 will comment on this enlarged perspective unifying FS and the universal viewpoint.

[28] "McNelis reaches for doctrines that are ridiculous in present cultures … [and that will be] mistaken for being commonsense talk that is way off the realist track. It will, perhaps, be centuries before talk of his mistaken optimism will be seen as the commonsense-nonsense expression of a brutal truncation." Email of Phil McShane. Here's an indication of the marketing challenge and one reason for the earlier claim that FS8 may be the most difficult of the specialties.

products to diverse audiences. Those backgrounds will be assets in communicating with broader audiences. Thus, some subset (an "auxiliary") of the practitioners of FS8 will be better suited than the rest for this task.

What is to be handed on? Given the focus of this conference, we want to reorient Lonergan students so that they take seriously his "third way." The first message passed along is negative. While Lonergan's work showed promise for remedying many cultural ills, Lonerganism has failed to demonstrate and make credible that promise among populations beyond the "Lonergan circle." This failure should be an embarrassment and an explicit topic of frank discussion. After so many years of conferences, articles and books, why is there so little to show? Reasons and excuses may be plentiful, but might the basic flaws have been the neglect of FS and the reluctance to take seriously the contemporary demand for serious understanding? To continue with the routines of the Lonergan circle doesn't seem a smart move; however, a familiar good of order, no matter how defective, offers predictability and security. Thus, the negative message has to be followed by persuasive evidence that something "strange" is worth trying.

Again, evidence or proof of effectiveness will lie in the doing of FS. Perhaps going "outside" the Lonergan circle in doing FS is one way of assembling evidence. The earlier reference to an "inductive" approach bears repeating. Functional collaboration is already occurring in some fields. Reporting on increases there in efficiency and on instances of cumulative results would be one way of amassing evidence. In addition, proof that some have succeeded in reaching audiences beyond the Lonergan circle may persuade more of his students to venture farther from home. Bob Henman's recent article is an example of reaching out to audiences in the neurosciences.[29] Sean McNelis has similarly been addressing audiences interested in housing policy.[30]

What are other strategies for handing on what we've learned about FS? How at least can we make it a more common topic of conversations within the circle? My stance is that we should target new generations of students, those not yet wedded to a single line of thought or too busy with academic careers. Undergraduate and graduate classrooms can be sites of "rebellion" against the inertia of the academy.[31] For example, students in the Humanities are used to what-questions, but both they

---

[29] Robert Henman, "Can Brain Scanning and Imaging techniques contribute to a theory of thinking?" http://www.crossingdialogues.com/current_issue.htm vol. 6, issue 2, December 2013. Since I wrote this there have been two further articles of his in that journal, and now all three are available in a larger book, Robert Henman, *Global Collaboration: Neuroscience as Paradigmatic* (Vancouver: Axial Publishing, 2016).

[30] Assembling further examples of such outreach may emerge as one of the outcomes of our conference.

[31] In the final section, my remarks on an adequate historical perspective are one indication of how one might reorient younger audiences.

and most of their instructors are unfamiliar with the limits of nominal understanding and the demand of further why-questions for explanatory answers. Exposed to appropriate examples, some of them will notice an "existential gap" in themselves, their textbooks and their teachers. A bright subset of them may, as a result, have a better chance of being "ruined" for life in all the old familiar places and may search for something different.

The really tough question about handing on is the "structural" question. How is FS8 to be part of a set-up, a good of order, one that maintains a flow of good ideas, experiments and records of successes and failures? First is the problem of "critical mass." Until sufficient numbers of specialists are at work, there are not many materials to build upon.[32] And until those members and materials are realities, there is nothing flowing. The old puzzle of which comes first, individuals or institutions, is not the issue. As a practical matter, loose forms of organization such as SGEME can "get the show on the road" as long as enough members regularly communicate what they are working on and for what projects they are seeking collaborators. Perhaps the main challenge posed by insufficient numbers is that not enough specialists will be interested in the same topic. As a result, dividing up the labor in pursuing a common project will not occur.[33]

So perhaps the question of handing on takes second place to the question of filling in what is lacking, namely, a series of experiments in the eight specialties focused on a shared problem. The conference has a shared focus, but the diversity of papers indicates a less than sustained focus on one problem. The practical question, then, is: Can enough of us agree upon a common topic and pursue it according to our talents? Difficulty in answering this in the affirmative probably returns us to the question of critical mass.

So what are we to do? Again, if identification is performance, we need to perform, even if our initial efforts are amateurish and disjointed. Despite that they still can be productive. We will learn by doing, by beginning to assemble materials

---

[32] I experienced this problem in writing two books on comparative interpretation. Doing FS4 presupposes the prior work of other specialists, especially the genetic histories of FS3. In the absence of those materials, I fabricated a few developmental histories or made use of "thin" versions already available in regard to some intellectual debate.

[33] For example, Bob Henman and I have been researching and writing about various problems in the neurosciences. An "anomaly" in the recent literature has been an explicit recognition of deficits in the language regularly occurring in that literature, e.g., talk of "mechanisms in the brain." Major figures have voiced dissatisfaction with "bottom up" accounts of mental acts but have also complained they lack an adequate vocabulary for "top down" accounts that would supplement the former. I see this as an opportunity both to reach beyond the Lonergan circle and to experiment with FS in relation to a serious issue in a serious set of sciences. However, are there enough hands-on-deck to get the ship underway?

for others to use in their specialties and by providing some evidence that FS, even at a very immature stage, can produce works superior to conventional academic fare. Being enthusiastic about such initial efforts will perhaps depend on what we think about history and our capacities to affect its course. The final section offers some general remarks on why I think even early ventures in the third way are worth our time.

IV. Credibility and Hope

The meaning of "existential gap" is complex. At a minimum it means the disparity between one's horizon and the "field" that is all of being. More specifically, it means a refusal or failure to seek comprehensive understanding by operating in the second horizon. But an implication of the latter usage is that a gap exists when one neglects questions about history and one's place in it. What is history the history of? Do we expect that it might ultimately be an intelligible order? How might that expectation be reconcilable with the nonsense of the present and past? What roles might we play in lessening any unintelligibility during our life spans? How willing are we to make the effort?

If our goals are implementing metaphysics and so changing conventional mindsets and practices, how credible will our initial ventures be? Should we hope for much? Perhaps we need a more adequate realism. What might a more adequate realism, a more adequate historical perspective, be? Think of Pat Brown's envisaging of exponential growth. Then add: You and I are extensions and durations "belonging to a manifold governed by emergent probability." What is "the field or matter or potency in which emergent probability is the immanent form of intelligibility"? We are among all those things that as "concrete extensions or concrete durations are the field."[34] Next broaden the perspective even more. For us "the highest level of integration is, not a static system, nor some dynamic system but a variable manifold of dynamic systems" open to ever further development.[35] Finally, suppose that some part of that variable manifold of dynamic systems is a community of scholars and scientists bound together by shared understanding and judgments, oriented toward communicating the good in human history and sustained by personal relations. Does FS now come into view as a manifold of collaborating unities, a community of inquiry with the promise of eventual effectiveness?

From this perspective, talk of the genetics of the genetic accounts of meanings seems quite realistic. What is received, refined and passed along in functional collaboration is to be fit into a developing ordering of views. What authors say about

---

[34] *Insight, CWL* 3, 195.
[35] *Ibid.*, 532.

a particular issue is related to a history of constructed meanings about that issue, and that history, in turn, is related to a further developing view of the meaning of history.

A change in perspective on one's own life may help here. "Your personal project and production may be no great shakes, no great success in neighbors' or historians' eyes; but it is a hidden unique loveliness savored fully only by God."[36]

To conclude, effective strategies of communication will take the labors of generations of FS8 specialists and far lengthier works than this volume of essays. But a gathering of the willing is a start, and the willing are helped along by intimations of the broader scheme in which they are invited to collaborate.

> Finally, good will is joyful. For it is love of God above all and in all, and love is joy. Its repentance and sorrow regard the past. Its present sacrifices look to the future. It is at one with the universe in being in love with God, and it shares its dynamic resilience and expectancy. As emergent probability, it ever rises above past achievement. As genetic process, it develops generic potentiality to its specific perfection. As dialectic, it overcomes evil both by meeting it with good and by using it to reinforce the good. But good will wills the order of the universe, and so it wills with that order's dynamic joy and zeal.[37]

---

[36] McShane in *Process*, quoted in *The Everlasting Joy of Being Human* (Vancouver: Axial Publishing, 2016), 73.
[37] *Insight*, *CWL* 3, 722.

Philip McShane

> "Look!
> Hidden beneath your feet
> Is a Luminous Stage
> Where we are meant to rehearse
> Our Eternal Dance!"[1]

The format of my essay is that imposed gently by me on most of the other contributors: two middle sections struggling for functional talk that were to be bracketed by two sections of musing helpfully about our efforts. The form enabled us to make identifiable efforts at functional writing, part of the identification being our own bracketing reflections.

I. Contexts

My first struggle with communications, in the context of Lonergan's writings, was in 1957, when the problem of Cosmopolis was raised for me by chapter seven of *Insight*. But it was almost a decade before it was lifted into a serious problematic context through the hand-on nudgings of Lonergan in an afternoon of the summer of 1966. After that nudging I paced the fields and indeed the *Field*[2] seeking light on the turn from the past to the future that was weaved into a dancing vision by Lonergan in February of the previous year. And it is worth noting that, prior to that nudge, I had the great advantage of a decade's musing on the problem, a decade that tuned me to his search for an answer to the problem of effectively changing global culture. His acorn answer slowly blossomed in the Field into a tree that transformed the turn into a vortex,[3] a vortex within each subject that promised to weave all subjects round a surreal[4] scary caring tower of subjects.

---

[1] I quote from the poem "Someone Who Can Kiss God" by Hafiz. I do so from Daniel Ladinsky's *I heard God Laughing: Renderings of Hafiz* (Walnut Creek, CA: Sufism Reoriented, 1996), 25. (hereafter *Hafiz*). Hafiz was born in Shiraz, in southern Persia, and lived most of his life there. His dates are probably 1320–1389, later then than Rumi (who died on December 17, 1273: interesting dates).

[2] "The field is *the* universe, but my horizon defines *my* universe," Bernard Lonergan, *Phenomenology and Logic*, *CWL* 18, 199.

[3] See note 10 below.

[4] See note 24 on page 24 of *The Everlasting Joy of Being Human* (Vancouver: Axial Publishing, 2013) on Lonergan's position as surreal.

I write in Proustian poise, freshly sensing finitude: might you come with me thus? : It's a piece of cake.[5]

In the summer of 1969, looking at the few shelves on music in the Old Bodleian library in Oxford, those shelves of God communicated to me the needs of musicology, and I weaved the communication into a medley lauded at the Florida Lonergan conference of Easter 1970 and as quickly forgotten as the *Gregorianum* article of 1969.[6] The four background chapters tiredly added by Lonergan to that article did not help memories or mindings, indeed they helped us enthusiasts who invented the *Annual Lonergan Workshop at Boston College* after Florida to stay off-track well into the next century.[7]

It was not until the early 1980s that I began to struggle with some competence round about the fourteenth chapter of *Method in Theology*. Then I began to sense the mastery of its first two sections.[8] But only in 2013 did I begin to reach the present view of the failed chapter and its brilliant seeding of an ethos of global communication.[9]

Previous to that there was the steady climbing of my eighth decade, beginning from a meeting with *The Canto*'s of that oddity Ezra Pound.[10] The decade of the *Cantowers* was, indeed, a climb of staggering discoveries, publication rejections and failed communications too complex to muse over here, but I would draw attention to the long essay of May 2003, *Cantower* 14, "Communications and Ever-ready

---

[5] Proust readers will recognize the ambiguity of the phrase. The piece of cake was the preoccupation of a life-time. The issue of adult growth raised by Proust haunts the present paper. My first serious thematic of it was in the concluding pages of *Lack in the Beingstalk* (Axial Publishing, 2006).

[6] "Functional Specialization," *Gregorianum* 50 (1969), 485–505.

[7] The sort of stuff we were at dictated the structure of the 2004 centennial gathering in Toronto. The gathering drove me to a week's musing on the situation, a musing that led to my *Quodlibet* 8, "The Dialectic of My Town, *Ma Vlast*," available at: http://www.philipmcshane.org/quodlibets.

[8] The essay of 1985, "Systematics, Communications, Actual Contexts" (*Lonergan Workshop*, volume 6, 1986), contains a suggestive question about the first section, titled "Ontology and Meaning. "Could it be read profitably under the alternate title, 'passionate subjectivity in the lucid closed options of the finality of implementation'?" The essay is now available as the seventh chapter of my book *ChrIst in History*, available at: http://www.philipmcshane.org/website-books. The suggestive question is on page 5.

[9] See my *Futurology Express* (Vancouver: Axial Publishing, 2013), and the Epilogue to *The Everlasting Joy of Being Human* (Vancouver: Axial Publishing, 2013).

[10] From Pound—backed by Flaubert's *La Spirale*—I rose to the notion of vortex. "If you clap a strong magnet beneath a plateful of iron filings, the energies of the magnet will proceed to organize form. ... The design in the magnetized iron filings expresses a confluence of energy." Ezra Pound, "Affirmations, Vorticism," *The New Age*, xvi, 11, January 14, 1915, 277. I quoted this text in *Cantower* 1, at note 39.

Founders," where I reached for a fresh meaning of Lonergan's two field-trips, *Insight* chapter 14 and *Method* chapter 14. That effort grew slowly through the next decade into my field-flight, in the summer of 2013, of *Futurology Express*, chapter 14, "Structuring Systems in Towns, Gowns and Clowns." On the road I had suggested a relocation of sections 3–5 of *Method in Theology* chapter 14 in chapter 1 of the book, but what was to replace them in chapter 14? The core of an answer to that question is the topic of my contribution here to the problem of instituting and fostering "the cumulative and progressive results"[11] that are to be expected, with normal-curve statistics,[12] from global omnidisciplinary collaboration in later millennia.

The question that haunts all of us that have gathered to live into and beyond the Vancouver Conference of 2014 is, **HOW**[13] to bring about a break forward in a stale Lonerganism that dodges his crazy invitation to a new tradition of contemplation. The elements of future meaning are the local layered mansions of collaborative groups.[14] Those layered mansions are within the ancient molecules of all of us. That skyscraper is emerging as a need more visibly in human housing, in salvaging ecologies, in ethnomusicology, than the usual elements of meaning sloganized by a decadent school.[15] So,

> one stumbles upon Hegel's insight that the full objectification of the human spirit is the history of the human race. It is in the sum of the products of common sense and common nonsense, of the sciences and the philosophies, of moralities and religions, of social orders and cultural

---

[11] *Method in Theology*, 4, 5.

[12] There is a solution to, and a statistics of, the problem of evil that Lonergan raised in chapter 20 of *Insight*, but it is, even heuristically, enormously complex. You might think of it popularly and existentially in terms of the leading question of the eighth chapter of *The Everlasting Joy of Being Human*: "Do you view humanity as possibly maturing—in some serious way—or just messing along between good and evil, whatever you think they are?" *Everlasting Joy*, 77.

[13] I boldface this word to draw attention to a central issue of these next millennia—the development of the linguistic feedback of a **HOW** talking, where the **H**ome **O**f **W**onder becomes radiant in every face, in every phrase. See note 24 for an intimation of the fullest Christian contexts, relating to vestiges of the Trinity.

[14] The sentence is an expression of the core of the key insight. It is quite obviously an inadequate expression, and points to the inevitable flaw in my brief expression. To those in the know, it is reasonably adequate, but for most there is the tough climb, best with help, through varieties of diagrams and illustrations. It is a topic for the larger work I mention in the concluding pages here.

[15] I have written abundantly regarding the extraordinary oversighting of the "What-to-do?" question in normal Lonerganesque talk. An initial help is Appendix A of *Phenomenology and Logic*, CWL 18, 322–23. The thesis of the priority of functionality over consciousness-identification is at the heart of chapter 1 of *Method in Theology: Revisions and Implementations*, available at: http://www.philipmcshane.org/website-books.

achievements, that there is mediated, set before us in a mirror in which we can behold, the originating principle of human aspiration and human attainment and failure.[16]

Metaphysics is, not the preserve of clowns in self-preserving departments, but a matter of the human story making globally luminous the mountain mirrors of dreaming ancient hills and tireless waves. We carry those dreams inside, in zealous but frustrated molecular patterns. Lonergan's *Field of Dreams*[17] eludes the main drive of his present disciples.[18] The old warrior found the elements of cumulative progress late in life. They are Gospel Fruit, binding cords and chords of the Symphony of Jesus.[19] The disciples, in the main, find only nots and bolt blindly from the chords in a crazy busy deafness.

So:

"You don't have to act crazy anymore —
  We all know that you are good at that.

      Now retire, my dear,
  From all the hard work you do

Of bringing pain to your sweet eyes and heart.

      Look in a clear mountain mirror —
    See the Beautiful Ancient Warrior
        And the Divine elements
        You always carry inside"[20]

---

[16] I quote from page 5 (header, page 14) of a Lonergan archival file labeled A697. It contains a typescript numbered from 8 to 23. Very plausibly it is a continuation of the sketch, from early 1965, of a first chapter on *Method* to be found in what I named—in 1974—V.7. This file contains nine pages of typescript that is pretty evidently a shot at a first chapter, and there are four handwritten pages there towards an entire chapter.
[17] I am recalling the title and content of the 1989 film adaptation of W.P. Kinsella's novel *Shoeless Joe*. Is there a parallel? Certainly there is a neglected baseball diagram right there in the economics of the Dream-Tower. (See the reference in the next note, page 163, "The Tower of Able: Lonergan's Dream.")
[18] I have written of the unshared context of Lonergan's heart-drive in chapter 10 of Pierrot Lambert and Philip McShane, *Bernard Lonergan: His Life and Leading Ideas* (Vancouver: Axial Publishing, 2010).
[19] The heuristics of the Mystical Body, the Symphony of Jesus, is the central topic of my *Method in Theology 101 AD 9011: The Road to Religious Reality* (Vancouver: Axial Publishing, 2011).
[20] *Hafiz*, 5.

II. Contents

Foundations are persons poised in kataphatic fantasy and cycling-dynamics. The poise sublates the anaphatic fantasy and dynamic nudging of poets and dancers, peasants and mystics, of which I take Hafiz as a convenient exemplar.

Each makes their own, as stumblingly best they can, the foundational suggestions of Lonergan, linked here by me, fortuitously, with the versifying of Hafiz. So, the "critical method" [21] that ends the reflections on nescience of "eighthly"[22] at the end of chapter 19 of *Insight* sublates incarnately the molecular shiftings in Hafiz invitation:

"If you think that the Truth can be known
From words,
If you think that the Sun and the Ocean
Can pass through that tiny opening
Called the mouth,
O someone should start laughing!
Someone should start wildly Laughing –
Now!"[23]

Now?

It is both the elusive divine Now and
The now that is called
A *sacrament*
Of the present moment or
The moment in the rose garden.
But for the foundation person it is
A day and daydream moment of a mind
Minding luminously W₃,
"Double You Three,"
In a habitual reaching-resting,
"Double You Three
In me,
In all,
Clasping,
Cherishing,
Cauling,

---

[21] *Insight*, *CWL* 3, 708: the final page of chapter 19.
[22] *Ibid.*, 705. There is the fuller context of thesis 5, *The Triune God: Doctrines*, *CWL* 11.
[23] *Hafiz*, 43.

Craving,
Christing."[24]

It is a minding contexted by
An up-to-dateness
Of globopolitical character,
listing forward in the foundational list
Of the nine familiar neglected bracketings
Of *Method in Theology* 286–87.

That up-to-datedness has its cutting edge
In the seeding symbolization that lurks in
A geohistorical imaging of the symphonic Jesus,
A haunting of foundational minding
In a yearning Ontology of Meaning.
The yearning,
A luminous dynamic of the full cyclic group,
Is a yearning for Common Meaning
In the Ontology of all and each situation of human beings,
All situations being
Piccolo-tunes
Within an Integral Symphony.

The foundational group
Of communications
Shares an imaging
Of each and all situations
That brings forth continually
The problems and possibilities
Of all particular situations.

---

[24] Perhaps a context from Hafiz would help your struggle here, wound round the struggle of Thomas. "I hear the voice / Of every creature and plant, / Every world and sun and galaxy – / Singing the Beloved's Name." *Hafiz*, 153. Add the hints of note 13 above. The Christian thinker has the problem of coming to grips with the beloved being Three and having more than three names: so, e.g., the name *Cauling* or *Calling* belongs to the traditional first and second Persons. We are in the world here of the puzzling of Thomas and Augustine regarding vestiges of the Trinity (see *Summa Theologica*, q. 45, a. 8). We are in the problematic of the emergence of a kataphatic praying desperately needed in this millennium. See the five essays *Humus* 4–8, on "Foundational Prayer," available at: http://www.philipmcshane.org/humus. See, further, James Duffy's contribution to the present volume.

It shares in a new renewing reading of *situation*
As it occurs at the end of Lonergan's consideration
Of "Common Meaning and Ontology."

What is that shared meaning,
Symbol-stretched
Spirit-sprung?

Each situation needs hovering over it –
Or should I not write towering over it? –
An eight-storied tower
Of separate situation-rooms.
These rooms all-round attend all-round
To the situations that are the topics, the places, of the eighth specialty:
An institute of government or crime,
A city block,
A campus,
A classroom,
A bank,
A temple,
A bedroom,
The seat on which you sit
Now.

Hovering over each in sacred global care
Is to be a strange topology
Of distinct mansions of meaning,
Topped by the caring research group
That notes what needs present care
In the being
And the well-being
Of that
Absolutely Supernatural
Particular situation
As it is
Thus superglued
Into finitude's glory.

Is to be?

There is the tragedy of a failed introduction.

"Hafiz introduces himself as Companion and Guide, Friend and Lover. He invites us to share his life, his wine and his heart, to see ourselves and the world through his eyes. If we didn't know better, we would think he was courting us—and perhaps he is!"[25]

We have been courted in vain by Lonergan.

III. Hand-on Strategies

The hand-on strategies relate to the single key insight of the content, and this is paradigmatic of the cyclic process, a paradigm taken from normal successful sciences, especially from work in mathematics and physics. But the paradigm wilts here in the face of the larger problematic of the failed introduction mentioned at the conclusion of the previous section.

I tackle the hand-on here in an apparently simple manner, but the readership varies in a manner foreign to successful science[26] and so lurking within that simplicity are hidden layers of difficulty of which I write in the bracketing sections one and four. Might I have continued with the sort-of versification technique that I used in the previous section? The possibility and the temptation were there, but here simplicity is desirable in order to indicate less discomfortingly the normal scientific challenge as best I briefly can under the present muddle of circumstances.

So my hand-on deals with the key insight in terms of two lay-outs of the task and achievement of meeting the needs of concrete situations. The first lay-out is that talked of by Lonergan in chapter 14 of *Insight*. How are these needs to be met? There is an illusory simplicity of three layers stated here in a convenient or contorted fashion: major premise of methods, minor premise of sciences, conclusions in and of concrete situations. A light-weight reader of *Insight* gets the point easily enough. A more serious reader, thinking concretely in terms of the groups involved and of the problem of Cosmopolis, may see the triple-layering as simply a strategic hiding of that problem. The up-to-date readers will see the key insight as a lift of their previous heuristic, a lift that pivots on a powerful local imaging, a lift, moreover, luminous—because of their antecedent heuristic—in and about their own needed molecular changes within the envisagement.

But initially, for all, the minding and molecular changes are merely a nominal hope. The key insight is simply expressed—"an invitation to see ourselves and the

---

[25] *Hafiz*, 19.
[26] By successful science I am excluding the growing flow of popular presentations. That tradition gained a huge lift from Fontenelle. There is a sense in which theology and philosophy, in the main, never rose out of that flow. On *haute vulgarization* and its illusions, see Bernard Lonergan, *Philosophical and Theological Papers 1958–1964*, *CWL* 6, 121, 151.

world ... courting us"—as seeing now, smelling now, each city block or rural farm as under the umbrella of an eight-layered towering collaboration of situations. Each situation is a group of functional collaborators but with a massively curious topology, a topology the diagraming of which is left now to your patient climbing global doodling.

How is this imaging and its heuristic content to be effectively handed on? On the analogy with successful sciences my skimpy paper enters the cycling of progress, beginning with the group of collaborators in this conference and volume.[27] The group of collaborators, at present an annoyance to Lonerganism, will ferment forward into various rewritings of the demands of the fourteenth chapter of *Method in Theology*. The embarrassing doctrinal shift will become increasingly effective in bringing about the death of the prevailing decadence in Lonergan studies.

But only if the initial group, increasingly living in an effective "apprehension of the group's origin and story,"[28] gives rise in itself and in a growing group of disenchanted Lonerganists, to an effective sensitivity regarding and guarding how Lonergan "invites us to share his life, his wine and his heart."

At age thirty he wrote, desperately, "what is to be done? I have done all that can be done in spare time and without special opportunities. . . . Briefly the question is: shall the matter be left to providence to solve according to its own plan, or do you consider that providence intends to use my superiors as conscious agents in the furtherance of what it has already done?"[29] Thirty years later the old warrior, battered by a life inflicted on him by a mindless religious culture, did not have the energy to write of the complex demands of effective implementation. That mindless culture now putters along around his poetry as a "MuzzleHim Brotherhood."[30] The group of collaborators in this volume, and those they attract, have their own mindless superiors and cultures and monsters to face: might some few do that "with indomitable courage"?[31]

---

[27] Of course, I like to think of the group and its allies as part of that "not numerous center, big enough to be at home in both the old and the new, painstaking enough to work out one by one the transitions to be made, strong enough to refuse half measures and insist on complete solutions even through it has to wait." "Dimensions of Meaning," *Collection, CWL* 4, 245.

[28] Lonergan, *Topics in Education, CWL* 10, 230.

[29] I am quoting here from the conclusion of a long letter of Lonergan to a superior, written in 1935. The letter is reproduced in full in *Bernard Lonergan: His Life and Leading Ideas*, 144–154.

[30] This is the title of my gloriously offensive chapter 8 of *The Everlasting Joy of Being Human*.

[31] I am recalling the Frontispiece of my *The Shaping of the Foundations* (1976), available at: http://www.philipmcshane.org/published-books, from Gaston Bachelard's *The Poetics of Space* (Boston: Beacon Press, 1969), 61: "Late in life, with indomitable courage, we continue to say that we are going to do what we have not yet done: we are going to build a house." It has been a long continuation since then of unheeded saying. It reminds me of my loose

Or are we to be boxed into the dodging of the Beloved's historical invitation?[32] Surely you must halt, now, Now, nownow, to read in "startling strangeness"[33] those five simple words of the 65-year-old master, "**it is a major concern**"[34]: that hands-on reach from his typing hands on to paragraph, parashoot, you into thinking of the Invisible Tower of situations that should hover over your sitting and your psyche.

"Once I asked my Master,
'What is the difference
Between you and me?'

And he replied,
'Hafiz, only this:

If a herd of wild buffalo
Broke into our house
And knocked over
Our empty begging bowls,
Not a drop would spill from yours.

But there is Something Invisible
That God has placed in mine.

If that spilled from my bowl,
It could drown this whole world.'"[35]

IV. Further Contexts

Why, you may ask, do I add to my poise the poise of poetry? But the issue is your poise, as you sit, now. My audience continues to be the vague group of readers that spreads out randomly beyond my compatriots in this "crisis."[36] For these compatriots, my poetic positioning is not an addition, for they know and feel the struggle

---

memory of a saying of Pablo Casals when conducting once a final rehearsal, "I have been trying to get this right for forty years: maybe tonight?"

[32] The Prologue to *The Everlasting Joy of Being Human* deals with the unaccepted invitations of both Thomas and Lonergan.

[33] *Insight, CWL* 3, 22.

[34] *Method in Theology*, 355. Bold-facing mine.

[35] *Hafiz*, 95.

[36] I refer here to the familiar text on aesthetic apprehensiveness during a crisis: *Topics in Education, CWL* 10, 230.

FOUNDATIONS OF COMMUNICATIONS

for their own integral luminous consciousness. [37] Nor is the weave of aesthetic reference foreign to them. Among my strange capers familiar to them is that other "14," the fourteenth chapter of *Lonergan's Standard Model of Effective Global Inquiry*,[38] where I weave the thirteen songs of Sinead O'Connor's CD, *Faith and Courage*, into my thirteen-sectioned appeal for an earlier skinned-knee view of the Field of Dreams. That caper gives us a further initial context here in so far as you might fantasize what might emerge from weaving sixty reflections round the sixty songs of Hafiz on the compact disk edited by Ladinski.[39] Might it not be "the inception of a far larger work"?[40]

But not mine.

Like Sinead O'Connor, and also like you—are we not back at the end of the first section?

"I have a universe inside me
Where I can go and spirit guide me
Then I can ask oh any question."[41]

And we can ask oh any question, but now strategically within our little mansion room in the "Dark Tower."[42] The questions are to ferment out of "the monster that has stood forth in our day"[43] including the monsters of Aristotelianism, Thomism, and Lonerganism and all the other effete ungrounded "academic disciplines" that so easily sit with the idiocies of the military-industrial complex, of financial murkiness, and of religious brutalities.

---

[37] A context is *Bridgepoise* 3 and *Bridgepoise* 10, a two part essay on "Liberal Arts: the Core of Future Science," available at: http://www.philipmcshane.org/bridgepoise.
[38] The book is available at: http://www.philipmcshane.org/website-books. Chapter 14, titled "Communications: An Outreach to Lonergan Students," is a 90-page final chapter to the book, with 13 sections corresponding to the 13 songs on the CD of Sinead O'Connor, *Faith and Courage*, Warner/Chappell Music Ltd, 2000.
[39] It is not, of course, a compact disc: but a flight of fancy can reach to the poetry's regular expression in the musical patterns and rhythms of the *ghazal*, the love song.
[40] *Insight*, 754.
[41] From the first song of Sinead O'Connor's *Faith and Courage* CD.
[42] *Cantower* 4, "Molecules of Description and Explanation," (available at: http://www.philipmcshane.org/cantowers) is a context here, with its explanatory—Tomega Principle—and feminist lift of Robert Browning's *Childe Harold to the Dark Tower Came*. But now I talk of a more complex Tower, with the present imaging conflicting creatively with the usual image in W₃ of the Tower of Able. See *Bernard Lonergan: His Life and Leading Ideas*, 161 and 163.
[43] *Method in Theology*, 40.

"Out of history we have come
With great hatred and a little room."[44]

And perhaps even a little barroom, with "A Barroom View of Love"?[45]

"I would not want all my words
To parade around this world
In pretty costumes,

So I will tell you something
Of the Barroom view of Love.

Love is grabbing hold of the Great Lion's mane
And wrestling and rolling deep into Existence

While the Beloved gets rough
And begins to maul you alive.

True Love, my dear,
Is putting an ironclad grip upon

The sore, swollen balls
Of a Divine Rogue Elephant

And
Not having the good fortune to Die!"[46]

---

[44] From the tenth song of Sinead O'Connor's *Faith and Courage* CD.
[45] The title of my final poem from Hafiz.
[46] *Hafiz*, 79.

Michael Shute

My task has been eased by the work of others in this volume. Generally, section 1 of Terry Quinn's "Interpreting Lonergan's Fifth Chapter of *Insight*" provides a concise, overarching context.[1] More specifically, Bill Zanardi's "Identifying the Eighth Functional Specialty" concerns the same functional field of communications as this essay.[2] I concur with Zanardi's account of the challenges and promises of *that* specialty at *this* time as well as his recognition of both our present efforts as a species of FS8>C9 (C89). What Zanardi has to say about communicating functional collaboration can be applied *carte blanche* to the problem of communicating Lonergan's economics and shortens considerably the work of Section 2 below (Contexts). In Section 3 (Handing on 1, 2, 3…) I provide three instances of FS8>C9 communications. Following the common pattern of all the essays in this volume, in Section 4 I include a short response.

As I understand it, we are at stage of development that is *prior to* a scientific revolution in general method[3] and, as it applies specifically to this essay, in economics. I say *prior to* because, I would concur with others in this volume that the emergence of functional specialization and macroeconomic dynamics in the mind and expression of Bernard Lonergan is a specification of the elements of a general collaborative method and the real variables for a science of economics. An operative science is an achievement of communal meaning and that only happens when there is genuine collaboration. To those familiar with the struggles to implement the general method of functional collaboration, it remains still, in the main, something to be hoped-for. There is as yet no accepted standard model in economics and as such the field of economics is pre-scientific. Lonergan's macrodynamic economics does specify the key variables for a standard model, but they are not common coin, as it were, in the field of economics. Practically speaking, in both the mainstream and heterodox community of economists they are not presently under consideration. So, it is this delayed revolution in both general method and in the field of economics that is my primary concern, a concern also addressed in other essays of this volume. This situation means there will be a delayed entry into functional communications (FS8) in the scientific field of economics. For these reasons, the most fruitful present strategy is to *speak* to various zones of common sense *listening*. In what follows I imagine a scenario where I can speak freely amongst functionally hopeful characters

---

[1] See Quinn's essay above.
[2] See Zanardi's essay above.
[3] See Philip McShane, "General Method," in METHOD: *Journal of Lonergan Studies* 13:1 (Spring 1995) 35–52, appearing later as chapter 5 of *A Brief History of Tongue: From Big Bang to Coloured Wholes* (Vancouver: Axial Publishing, 1999).

about strategies for communicating Lonergan's discovery of the real variables in economics to three kinds of common sense audience: (1) the world of economists, (2) the world of Lonergan scholars and students, and (3) everyone else.

I. Personal Context

And so I begin with my personal context. I was first introduced to the word 'functional specialization' in 1978. At the time I was involved in a local committee of the Canadian Catholic Organization for Development and Peace (CCODP). The group was committed to lending a very practical hand to solving the problem of world poverty. The people I worked with were good people who devoted time and energy to raising money and developing plans for raising public awareness of the plight of world poverty. After a year or so, however, I was becoming increasingly dissatisfied. In retrospect, the dissatisfaction was primarily with myself: while I was in accord with the good intentions of the group, I could not reconcile our approach with what I suspected about the economic situation. The problem was not to be solved simply through a redistribution of money, but I did not really know *why* I held this suspicion. Beyond that, I could not get a grip on what real economic development actually was. How was material wellbeing connected to cultural meaning and to the larger life of the spirit? I did not have a context for integrating what manifest itself as a set of seemingly disparate concerns. I discussed these issues with a fellow committee member and he suggested I read Lonergan. So I picked up *Method in Theology*. By the end of the first chapter I realized I had found elements of a solution to a fundamental methodological problem that had bothered me for over 10 years. The account of the structure of the human good in Chapter 2 provided hopes for an integrating context. Later I would discover the rich context of Chapter 15 of *Insight* that informs it.

After reading *Method in Theology*, I had a name, 'functional specialization,' that hinted at a way forward, but I had no solid notion of what the name meant. Much like Linnaeus' taxonomy of the plants, my understanding of the species of inquiry was descriptive: I could name the specialties and provide examples. I took a course on *Method in Theology* and was not much further ahead. I suspected that the meaning of the species-division was deeper and richer than what I was being taught about it. Thirty-two years later I was able to write an article called "Functional Collaboration as the Implementation of Lonergan's Method': For What Problem is Functional Collaboration the Solution?"[4] So I made some progress on that front. But in 1978 functional specialization or functional collaboration had not an *effective* meaning for

---

[4] "Functional Collaboration as the Implementation of 'Lonergan's Method' Part 1: For What Problem is Functional Collaboration the Solution?" *Divyadaan: Indian Journal of Philosophy and Education*," volume 24, No. 1 (2013).

me. I was still at the beginning stages of a development in my own understanding of scientific collaboration.

Initially my attention was directed to the preliminary and fundamental work of reading of *Insight* and working on the exercises in *Wealth of Self and Wealth of Nations*. Working on *Insight* and doing exercises in self-appropriation were sharply at odds with my formal education and, as it turns out, for the most part that dialectic reality remains the case even now as a thirty-year veteran in the education system. This situation is *a* context of the problem of the title. We have to communicate the form of a scientific breakthrough in the context of dysfunctional educational institutions.

Some years prior to my introduction to *Method in Theology* I had developed an interest in economics. As an undergraduate philosophy student I had read a lot of Marxist economic theory and later discovered Schumpeter's *Capitalism, Socialism, and Democracy*. Schumpeter cured me of any Marxist leanings and I enrolled in a graduate program in political science with the intention of writing a thesis on the method of political economy. That effort was a failure. I soon discovered that the methodology with which I was expected to work—a combination of Mill's utilitarianism and rational choice theory—was reductionist and in my judgment profoundly destructive. I lasted about fifteen minutes in an introductory macroeconomics course. The professor seemed doggedly committed to a utilitarian approach and I did not want to waste my time with something that was clearly (to me) flawed from the get go. I knew there was a serious problem with economic theory and that the problem affected most everyone. I would need something else to do with my dissatisfaction with CCODP. Later I picked up on McShane's focus on the economic question in *Wealth of Self and Wealth of Nations*.

However, it took me a while to get to Lonergan's economic manuscripts. It wasn't until I was working on my doctoral dissertation in 1984, the year of Lonergan's death, that I read the typescript of the 1944 version of *An Essay in Circulation Analysis*. Lonergan's approach was largely foreign to me. I had read Von Frisch, Darwin, and Mendel as a teenager and having worked as a naturalist specializing in plants I had the good fortune to have discovered the spirit of scientific inquiry. I appreciated Lonergan's genuinely scientific approach. Having read Marx and Schumpeter helped. While neither Marx nor Schumpeter managed to be fully dynamic, nonetheless, they knew an adequate economic theory had to be dynamic. Both in their own way grasped that there was a real connection between productive flows and monetary circulation. As I was working steadily away on *Insight*, and on myself as a dynamic flow, I grasped that Lonergan's methodological approach had great promise, though I could not at that point make a robust connection between the metaphysics of *Insight*, myself, and macroeconomics dynamics, let alone functional specialization. But certainly Lonergan's account of production as a transformation of the potentialities of nature, including our human nature, was for me a clear starting point. I started to grasp how my own transformation was also relevant to understanding economic flows. In 1984 I was, however, working on a thesis on the dialectic of history and so put off serious effort to work on the economic manuscript. I would later discover that the dialectic of history and macrodynamic economics were intrinsically connected and making that connection became a large part of the point of writing *Lonergan's Discovery of the Science of Economics*.[5] However, it wasn't until 1994 that I made a serious commitment to working on economics and worked up a grant project to work on Lonergan's economic manuscripts, which to my great surprise, was successful.

So what has all this have to do with the difficulty of communicating macroeconomic dynamics both functionally and effectively? I have been working on economics as time and disposition allow for twenty years. It has not been my only focus, but it certainly has inhabited my body and mind. From the beginning I tried to approach my own education in economics functionally. Even if they do not come anywhere near McShane's line-by-line criterion, the books I wrote were at least minimally divided into *functional research* and *functional interpretation*. *Lonergan's Early Economic Research* is an effort to 'make data available ' and *Lonergan's Discovery of the Science of Economics* an effort to interpret, in a genetic context, what Lonergan meant in his early and incomplete economic writings. I have also produced a number of efforts at communicating the distinction of two circuits. I judge these efforts failures insofar as they seem not to have had much influence in redirecting the flow of research on Lonergan's economics within the Lonergan community. I addressed this question at the 2009 Halifax gathering and came to the depressing conclusion that there was not much progress within the Lonergan community. At the time I counseled the long game, much as Fred Crowe regularly advised.[6] I am, however, beginning to wonder, given the current global economic mess and the damage that it does, who my audience really should be. So, keeping in mind the slow pace of history, is there a way of changing the odds or shifting the Poisson distribution for the emergence of economics as a genuinely collaborative science grounded in an appreciation of the basic variables discovered by Lonergan? Might this be better accomplished by expanding my audience and by shifting the style of communication? This is the question.

II. Contexts

In the current situation, to commit to working functionally with Lonergan's economic texts means you will have little company. This makes collaboration difficult other than the general realization that we 'live on the shoulders of giants.' As I suggested above, the theory itself is profoundly marginal in the public and professional discussion about economics and efforts to communicate about the

---

[5] Michael Shute, *Lonergan's Discovery of the Science of Economics* (Toronto: University of Toronto Press, 2010).
[6] See Patrick Brown's essay at note 6 and William Zanardi's essay at note 3.

economics both inside and outside of the world of Lonergan disciples and students' world has been ineffective. The tone and context of conference presentations among Lonerganians does not yet signal the real revolutionary shift of Lonergan's advance. I strongly suspect that for many 'Lonerganians,' the economics remains an oddity, an example of Lonergan throwing his hat into the *pastoral* ring.[7]

And this brings me to the occasion that sparked this paper. Recently I have been working through Lonergan's notes on economics from the 1970s and 1980s in conjunction with reading some post-Keynesian economists.[8] I have been trying to come to grips with the problem of understanding how a smooth transition from the surplus expansion to the basic expansion might happen and thinking about how I might then communicate such an understanding to heterodox economists.[9] In the course of my research, I noticed an interesting page from notes Lonergan made comparing Adolph Loewe and Robert Gordon, most likely made in preparation for teaching the 1979 version of his "Macroeconomics and the Dialectic of History" course.[10] What struck me in particular was a "?" on a page of notes. The question mark concerned naming the appropriate precept for the basic expansion. As we know (now there is an assumption!), for the surplus expansion there is the precept "thrift and enterprise" and Lonergan repeats this precept on the page in question.

---

[7] Typically such generalizations have exceptions and nuances. For instance, Gerard Whelan in his recent work *Redeeming History: Social Concern in Bernard Lonergan and Robert Doran*, *Analecta Gregoriana* 322 (Rome: Gregorian and Biblical Press, 2013) acknowledges the central role in Lonergan's own development occupied by the economics and the search for the method of functional specialization.

[8] In particular, George R. Feifel, *The Intellectual Capital of Michal Kalecki: A Study in Economic Theory and Policy* (Knoxville TN: The University of Tennessee Press, 1975), Steve Keen, *Debunking Economics: The Naked Emperor Dethroned?* (London: Zed Books, 2nd ed., 2011) and the monthly journal *Real World Economic Review*. I have also found periodically revisiting Joan Robinson's *Economic Heresies* (New York: Basic Books, 1971) very useful.

[9] The very difficulty of understanding *what* this sentence might mean is a very useful instance of the problem of this essay. It has some meaning for me related to understanding the acceleration in a surplus expansion at the point where in the context a pure cycle at the point where it is beginning to shift out of the surplus expansion, $dQ/Q'' > dQ'/Q$, and into the proportionate phase, $dQ''/Q'' = dQ'/Q$, in anticipation of a basic expansion, $dQ''/Q'' < dQ'/Q'$. But how to translate this appreciation in a form that communicates the meaning to a post-Keynesian economist who does not sharply differentiate the two circuits? And what about integrating the effect of the trade cycle in the analysis of the data under consideration? (See Lonergan's graph of in *CWL* 21, 274.) Now, to further approximate the actual macroeconomic situation in any phase at, say, the level of a national economy we need to add Lonergan's account of superposed circuits. *CWL* 21, 308–310. All roads seem to lead back to the statement 'there are two circuits' as the *sina qua non* for effective communications.

[10] See the Appendix: Between Gordon and Lowe.

Lonergan had also proposed a precept for the basic expansion stage in 1944, "benevolence and enterprise" but instead of repeating it here he put the question mark. Why? What about his reading and thinking in 1979 led him to mark the possibility of a reformulation of the precept? In the context of doing functional research, what significance or relevance does my picking up on the question mark have to the person working in functional interpretation? What am I passing on? And this would on the face it seem to be an appropriate context for this project on functional collaboration. The problem of the shift to the basic expansion can be found in the work of Kalecki (though not put that way) and those influenced by him. So this would seem to be an interesting place to make a stab at communicating to economists. Recall Lonergan saying to McShane in 1968, "Find me an economist!" But in fact to whom am I talking? Who is the economist? And is there really an audience among the post-Keynesians? Are they not also on the margins of the field of economics and in any case how empirically focused are they? And who is there among the Lonergan crowd? Perhaps a group of three, four, or maybe a few more? If that is the situation, then the question arises as to what purpose it serves to continue with my current focus on functional research in economics or even to carry on with the task of trying to figure out how to talk to post-Keynesians?

For the record, I do not think working through Lonergan's archival materials is a waste of my time. It terms of advancing in understanding Lonergan and economics, the time spent is valuable. However, in what sense is that part of a *functional* collaboration? And so it is the determination of the actual context for doing economics functionally that puzzles me. An advantage of functional collaboration is the efficiency of the division of labor it proposes. But how in fact do we collaborate effectively and efficiently in the current situation? By current situation I mean both the state of the research in Lonergan studies, the global "shambles of economics,"[11] and the stumbling efforts of getting 'functional collaboration' off the ground. What situation can we imagine where such collaboration is up and running? What is the critical mass for working collaboratively? What if the critical mass for collaboration is not present? How, then, do we *select* the projects we might work on, and even collaborate on in some meaning of the word?

In short, the situation is messy and "the messy situation is diagnosed differently by the divided community."[12] We do not have a standard model in economics or a standard collaborative method in academia. There needs to be a certain 'critical mass' of consent to the method of collaboration for the shift to begin. Both Quinn and Zanardi have addressed the difficulties of communicating functionally in our current situation where the set of functional communicators is null, or virtually so. In economics this is especially the case. As Lonergan noted in 1978, economics is the

[11] Philip McShane, *Futurology Express* (Vancouver: Axial Publishing, 2013), 5.
[12] *Method in Theology*, 358.

notorious instance of human science open to suspicion. [13] Thus the creative challenge is to develop strategies of effective communications in the real context we are presently in. We are still 'wandering in the desert' and while we might imagine the promised land of genuine functional collaboration in economics, the situation we encounter is more immediate and practical. For these reasons, I follow Zanardi in selecting the shift to FS8>C9 as strategically significant. As indicated above, I have identified three groups to speak to: (1) economists, (2) Lonergan disciples, (3) everyone else. In the next section I offer three such communications, *as is*. I will briefly reflect on these in section 4.

III. Handing On 1, 2, 3....

1. Talking to an Economist

I have talked to economists about Lonergan's economics but have been unsuccessful in getting to a point where there are some grounds for dialogue. No doubt there might be some ground for long-term optimism with heterodox economists, as Lonergan himself suspected but, fundamentally, there needs to be a minimal region of common ground. In my assessment without a possibility of dialogue on the reality of two circuits there is not much to talk about because my efforts to communicate Lonergan's economics will be interpreted into a context in which the insights will be effectively lost. The email that follows was my response to an article-length critical assessment of my book *Lonergan's Discovery of the Science of Economics* sent to me by an institutional economist who is a devote Catholic, sympathetic to Lonergan's theology. The main points of criticism of my account of Lonergan's approach concerned (1) the meaning of 'dynamic,' (2) my claim that Lonergan solved the problem of the quantity theory of money, and (3) the tone of my criticism of mainstream economics in the book. The paper was read at a Lonergan Workshop in 2011, and it preceded my own presentation on functional collaboration in the context of Catholic social justice efforts. I was informed that the critique would be given about ten minutes before the session started. If I had known it was coming, I would have offered a public response, which might have opened up the possibility for a debate.

Dear -------

Thank you for sending the full copy of your paper, which I have read carefully, and thank you for your detailed comments and criticisms.

---

[13] See Bernard Lonergan, "Moral Theology and the Human Sciences," in *Philosophical and Theological Papers 1965–1980*, ed. Robert Croken and Robert Doran, vol. 17, *Collected Works of Bernard Lonergan* (Toronto: University of Toronto Press, 2004), 302.

As a general disclaimer I did not write the book with economists in mind, but rather I had Lonergan scholars in mind, and that had a lot to do with my approach. I also had to learn what I could of the history of economic theory and to that end I relied on such texts as Mark Blaug's *Economic Theory in Retrospect,* Schumpeter's *History of Economic Analysis,* Rostow's *Theory of Economic Growth* as well as some of the important primary and secondary sources on Hayek, Frank Knight, Robinson, Kalecki, Keynes, the post-Keynesian economics and so on. Prior to taking up this project I had a beginner's familiarity with the field, but it was a steep learning curve that took up a fair bit of my time. There is the difficulty of learning a disciple as an autodidact and invariably there will be missteps. I am a theologian who main interest these days is in a theology of the mystical body! It is my hope that someone with a professional background in economics takes up the project of learning Lonergan's theory, as they will have a better notion of how to communicate with professional economists. I have been in touch with Paul Osslington in Australia and as I understand it, he is working on translating key discoveries into a mathematical language that macroeconomists can relate to. I certainly wish him all the best on that project. I will be happy to return to theology.

My main point is that economics is central to Lonergan's life work and not a peripheral concern and I hope that comes through. If Lonergan scholars do not take the economics seriously then I think it will more than likely fade away and I believe that would tragic. Even if elements of his writing appear or are dated, the fundamental discovery of two functional related circuits is crucial. This point needs to be communicated to the global community if we are to move forward towards a genuinely rooted reform of current economic practice. There are other elements I believe that are also important but the two circuits is key. Perhaps as important, there is the clear need for Catholic ethicists to take up economic theory seriously. On that point I am presently working on paper for *Theoforum.* I'll send it along for comments form you when I get a decent draft.

There are three specific issues I will comment on though.

First Dynamics. I believe you are correct in your surmise that Lonergan's notion of dynamics is something other than what it is in macroeconomics. I think it is central issue and I think the issue pertains both to economic theory per se and to general methodological issues that go beyond economics.

My assessment re: the claim that macroeconomics has not crossed the bridge into a full-fledged scientific dynamics depends in part on the reports of Blaug and Schumpeter. Blaug, for instance, writes in *Economic Theories in Retrospect:* "Mainstream economics treat dynamic analysis as a form of disequilibrium that takes into account temporal leads and lags in economic relationship" (211–12) In the same book he also says economic dynamics is for the most part comparative statics. Lonergan read Blaug's book on methodology in the 1970s. And Schumpeter wrote this in reference to Marshall and Walras and few others but in the context he meant the whole tradition of equilibrium economics: "By the phrase 'crossing the Rubicon,' I mean this: however, important those occasional excursions into sequence analysis may have been, they [economic theorists] left the main body of economic theory on the 'static' bank of the river; the thing to do is not supplement static theory by the booty brought back from these excursions but to replace it by a system of general economic dynamics into which statics would enter as a special case." (*History of Economic Analysis*, 1060) If things have moved forward significantly since Blaug and Schumpeter's assessments I would be ecstatic. But that move needs to be a full shift to a dynamic heuristic that takes with utter seriousness core methodological issues. I do not think that half measures will work on this point. As Lonergan once said and I quote loosely—you do not build half a ship.

Lonergan's notion of 'dynamic' intends to be dynamic in its core variables. So even when we are dealing with a stationary economy or static economy, it is dynamic. Behind this is his theory generalized emergent probability with its notion of a conditioned series of sequences of flexible schemes of recurrences. The theory of emergent probability is relevant to economics. That notion itself relies on his core discovery of cognitional foundations that are also fully dynamic. All of reality and all of history for Lonergan is dynamic and he spent much effort developing a general dynamics relevant across the board, whether we are talking about economics, philosophy of science or theology.

Second: As for my claim re: the quantity theory of money. I no doubt could have phrased my meaning better. My assessment however was based my own study of Philip McShane's essay on the subject from the appendix to *Past Keynes Past Modern Economics*. My thinking had less to do with current practice in economics as it had to do with Lonergan's own conception of the problem as we find it in the essay '*For a New Political Economy*' (1942) and in the set of documents he wrote between that essay and '*An Essay on Circulation Analysis*' in 1944 (Part 2 *Collected Works*, 21). If you study the transition between the two essays carefully, you find his work on turnover was central to his solving the problem for himself. I used Blaug's essay in the volume *The Quantity Theory of Money: From Locke to Keynes to Friedman* as a point of reference on quantity theory in the history of economic theory. For Blaug at least it is an unsolved problem.

The key is Lonergan's discovery is his work on turnover quantity and frequency, which I believe he understood as the full dynamic solution to understanding the function of money in a dynamic economy. It is not enough to count \$ but to set \$ in the context of turnover magnitude and frequency and taking in to account their accelerating affects. The claim may sound foreign to economists, as you say, but it makes sense to me if you are considering Lonergan's development, which is what I was focused on.

Finally: On the issue of the tone of communicating to economists: As a general rule, I think it a good role of thumb to assume and expect that the other parties in a conversation are open to understand what you are saying and conversely that you yourself be open to understand what they have to say: after all you may learn something. This is essential for any meaningful dialogue.

But besides dialogue there is dialectic. Sometimes the problem is that you cannot find a middle ground. My line in the sand is that there really exist two distinct and functionally related economic circuits. Either there are or there are not. If there are not, I am wrong and should be assigned to the dustbin of history. If there are two functionally related circuits, then the discussion has to start there. There really is no middle ground and to try to fabricate one is unreasonable: a science of any significance has to have a set of agreed upon fundamentals. This is an empirical question and it can be settled empirically. So there needs to be politeness, consideration yes, but also honesty. Truth matters.

But I wish to quote Lonergan here:

"If we are to escape a similar fate, [the dark ages] we must demand that two requirements are met. The first regards economic theorists; the second regards moral theorists. From economic theorists we have to demand, along with as many other types of analysis as they please a new and specific type that reveals how moral precepts have both a basis is economic process and so an effective application to it. From moral theorists we have to demand along with their own various forms of wisdom and prudence, specifically economic process and promote its proper functioning.

To put the point in negative terms, when physicists can think on the basis of indeterminacy, economists can think on the basis of freedom and acknowledge the relevance of morality. Again, when the system that is needed for our collective survival does not exist, then it is futile to excoriate what does exist while blissfully ignoring the task of constructing a technically viable economic system that can be put in its place.

Is my proposal utopian? It merely asks for creativity, for an interdisciplinary theory that at first will be denounced as absurd then it will be admitted to be true but obvious and insignificant and perhaps finally be regarded as so important that its adversaries will claim that they themselves discovered it."

Lonergan was the finest example of the authentic Catholic mindset, don't you think?

All the Best

Mike

I did not receive a reply to this effort. And therein lies one of the fundamental problems in communicating with even sympathetic economists. There is the long detour of dialectic that precedes dialogue. There are, of course, institutional problems, for example, what are the chances of getting a paper on two-circuit economics published in *The Economic Journal*?

2. Talking to a documentary filmmaker

This is a recent email exchange with a documentary filmmaker who has proposed doing a documentary on Lonergan. I had previously suggested that the economics was not peripheral to Lonergan's thought and had made a plea to include it in his plans for the documentary. I include his very respectful email and our subsequent correspondence.

2a. From Lowell to Mike

I'm going to keep poking here to see if I can bring about an insight in myself. Apologies in advance if my questions seem ignorant.

To begin. Yes, I understand that the two flows insight is central. However it still leaves us with such a simple answer on one hand, and almost no traction in the field of economics on the other.

So, on one level, I'm wondering whether the two-flow insight is the most fertile aspect of LB's work to enter into a discussion, or whether there is a better entry point. But let's assume it is...

... My first question would be, does current economics truly focus on one-flow only? Or is that an over-generalization. And more importantly, would an economist (lumping them all into one basket for the moment) think that there is only one main economic flow? And if the answer to the latter is true, then there is a real opening for discussion, but if the answer is no, where is the disconnect? Is it that all economists don't see their own assumption? Or it is something else.

Maybe this is reaching, but I'll relate two recent economic "differentiations" I've come across:

A. The first, in the book by Cooper that I referenced in my email to John. In one chapter, there was a specific point he makes about the difference between the dynamics in the market for products and the market for assets, and he notes that they behave differently (John, you cut and pasted a review from one of the 60 or so reviews of the book on Amazon, and in that review, this refers to his point 2—However, I believe the reviewer misrepresents the book which makes the distinction was not between goods and services, but between consumer goods and assets).

B. Likewise, with the statistical economist I spoke to, he pointed to a time series graph of returns, and noted the areas of huge variance compared to sections of small variance in the graph, and said that each component of the time series data needed to be handled differently, and teased apart to properly understand the data (and predict risk specifically).

Now maybe neither of these two examples is really connected to LB's two flows (other than the analogously of "just as such and such is handled through differentiated two aspects, so too is the general economy best handled by differentiating two flows..."). But where then does LB's insight about two flows couple with current economics in how current economics is done (not how it is explained, but how it is done)? Is it in how data is selected and collected? Is it in how that data is differentiated? Or even how events in a time series of data are differentiated (assuming the time series uses events rather than simply equidistant lumps of time)? Or is it a more fundamental difference in defining what an economy is or what its goal is?

I agree that one can convince a lay person—even a business person—of the "truth" of LB's two circuits and that this has been overlooked for a hundred years in economics, but that could well be because most people don't practice economics (in the way an economist does) and thus it all sounds very reasonable. But is it true. But in a similar manner I could explain nuclear decay to an interested person in a coffee shop, and they might nod and say they get it, but a physicist listening in from the next table might simultaneously be tearing her hair out at all the half-truths, over-generalizations and inaccuracies in my explanation. All to say, that I suspect there is more to it than entrenchment or bias on the part of economists. If it is such an easy insight to communicate, why hasn't it been taken up? Does the insight of two flows appear trivial to an economist and why? As a layperson, I find Mike's paper "Real Economic Variables" to be one of the very best communications of LB's economics I've ever read. But getting me to nod isn't the task at hand.

To risk circling back around, I'll quote form Mike's paper:

"Both GDP and GDI, which are standard measures of economic activity, lump together consumer and producer goods. Lonergan's claim is there is a real, functional distinction between the basic and surplus circuits that is fundamental for economic analysis. Without the distinction, there is no economic science."

And ask again… would a modern practicing economist agree with this statement? First, that this distinction is nowhere made. And second, that even if it is in some way, it is not made at a fundamental enough level to correct theory.

Lowell

2b. My response to Lowell

Thanks for your questions. I am very pleased you are interested in a doing a documentary on Lonergan. I once tried to pitch the idea to the CBC Ideas program—even had a producer who backed it—but got nowhere. I wish you better fortune than I had. It is a worthwhile project and, for what is it worth, in my view the economics is a core component of it. Lonergan work in economics is a concrete example of Christian living.

The way I would approach the economics is to understand it as a complete intelligibility, that is as a set of terms and relations where the terms fix the relations and the relations fix the terms all neatly brought together by Lonergan in the 'baseball diagram.' As you note, the 'two flows' are central to understanding the set of basic variables and, as I understand it, there is no way around communicating the breakthrough outside of the context of the two circuits or as Phil has put it "two kinds of firms." But beyond that, Lonergan also provides a precise account of the relationship between the flows: the surplus flow accelerates the basic flow as he formalized it in the equation on p. 244 of

CWL 21 (*CWL* 15, 37). He also clearly indicates the proper relation between the flows that should occur in any interval—the crossovers must balance. What goes from basic expenditures [E'] to surplus income [S''] in any interval is to be met by an equivalent flow from surplus expenditure [E''] to basic income [I']. If the demand for crossover balance is not met then you are draining one circuit in favour of another and there will be dysfunctional consequences. If you drain the Basic circuit, recession or worse; if you drain the surplus, then you cut off the surplus expansion (and likely add inflation). Lonergan also precisely differentiates re-distributional payments (Stock market, taxes, casino expenses and income, the real estate market) from operative payments that arise from the production of goods and services both in the basic and surplus circuits. He adds the complexity of transitional payments, recognizing that you only count the final sale. He adds in his account of the Pure Cycle, an explanation of how economic development can be achieved without producing a trade cycle, with its cycle of boom and bust, growth and recession.

Essentially you cannot understand one piece without understanding how it all holds together.

So no, it is not that the various components of Lonergan's theory have never been noticed except perhaps pure surplus income (Marx noticed what he called 'surplus value' but misunderstood its source and misapplied it). What is different is the way Lonergan has brought together systematically what others have treated partially or incidentally in the history of economic thought. Do other economics acknowledge more than one flow? Yes, some do. You can tease it out of Marx or Kalecki. Adolph Lowe would say there are even more than two (and in fact Lonergan would agree that there are higher flows that a simple surplus flow, but we need not get into that right now). Some economists (the noted liberal economist, Reich for instance) point to the difference between the "real economy and Wall Street." If we move into the pure cycle, elements of it are acknowledged when we recognize booms and recessions and these phenomena have certainly been recognized in the history of economic thought. But to quote Lonergan "if we succeed in working out a generalization of economic science, we cannot fail to create simultaneously a new approach to economic history." (*CWL* 21, 10) The real issue is that once you shift gears to Lonergan's approach you are in a whole new universe—the data are organized differently, the history is written differently, the issues that are noticed are different. To take an example from the history of Astronomy, once you move to a Copernicus' Sun centred astronomy, you will be hard pressed to put it in terms that Ptolemaic astronomers would understand or accept. And that is the problem. In my judgment, Lonergan's achievement is a discovery of a set the basic variables for a new science of economics. Once you move into it, you cannot go back. I have an understanding of economic flows that is in terms of two circuits, etc. I cannot cram that into past ways of doing the subject without the real point being missed altogether. It would like being asked to transform the elements of the periodic table into a debate about the merits of phlogiston. The periodic table makes the discussion of phlogiston moot. You have a whole ship; it only floats when all the parts are working together. If I were well schooled in equilibrium economics I could, as many have, pointed out its failures and mistakes. There is lots of those books out there and it not a new thing. Joan Robinson wrote her blunt *Economic Heresies* in 1971. Schumpeter's *Theory of Economic Development* was written in 1911. What Lonergan has to offer is a way to understand the mistake and correct it the real economy and it is as obvious as the nose on your face if you pay attention to what is actually going on in any business. However there is no traction because essentially you have to teach the new approach and it is very difficult to teach a professor :) because of the inertia of their habit as economists. To paraphrase Max Plank new ideas succeed when the old guys die off. So that all pretty depressing.

However, there is hope when you are speaking with fresh minds not schooled in mainstream theory. You can teach. I teach Philosophy of Religion in a Religious Studied department and I typically have a class with an even mixture of philosophy students and religious studies students. I want to get across GEM. After a year or two of philosophy, the philosophy students resist what I have to say, they keep arguing that I am being subjective. The religious studies minds usually are more open and they more readily enter into the puzzle of the self. That you the layperson are interested is more to the point than that the economics profession is convinced. Professional Economists are a somewhat like the Vatican Curia and they control access to the 'significant journals in their 'field.' Even Nobel Prize winner Paul Krugman has complained he cannot get published in them. http://krugman.blogs.nytimes.com/. Steve Keen, a heterodox economist, from Australia, argues quite persuasively that economics must undergo a long overdue intellectual revolution. (*De-Bunking Economics*, chapter 2 but passim). There is the Post-Autistic Economics movement http://www.paecon.net/that does organize dissent within the economics field itself.

The two examples you posed can be dealt with in Lonergan's economics. The way they would be handled would recast the inquiry in such a way that the questions would be radically altered. The talk would not be about 'assets and products,' but about flows of basic and surplus operative and redistributional transactions and in terms of the phases of the pure cycle and it its distortions. Time graph series etc. would be handled in terms of "conditioned series of recurrent schemes" and "probabilities schedules of emergence and survival." (*Insight*, Chapter 4) I would add that while Lonergan is empirical and scientific in his approach, most of economics is not done that way. So there is the further problem of dispelling the notion that economists currently are being scientific in their actual practice. Again not a new idea. See Joan Robinson or Schumpeter, above.

So, the short answer is, Lonergan's economics doesn't help much with problems the economics profession is concerned with. I could make a case but I am not optimistic that the solutions I offer would be heard. You can drag someone to therapy but if they are unwilling to do the necessary work then there will be no progress. I would like to be proved wrong and hope to find someone who has the formal training as an economist and is truly open to listening to what Lonergan had to say. If you find that person give them my email. Perhaps Eileen, who does have that training, can chime in here!

I watch and learn from documentaries. They are typically pitched to the intelligent person. Intelligent people can grasp a real problem as they can recognize a solution if it is well presented. The significant movements of our day are not going to follow the old routes. However, people are very interested in a solution to the economic mess we are in and it is a compelling mess that threatens us all. We need to jump on the right side of history. My hope is that we can be informed by the right theory and the right method. The question is how do we communicate it to the public so they are not deceived by bad theories, whether orthodox or revolutionary.

Lowell did respond, and believe it or not, the response came as I typed this sentence. Here it is:

Hello again Michael,

I took some time over the weekend to draw a schematic of how a working experiment might be created to look at monetary flows and collect relevant data. Like many things in life, it came about by combining two different projects occupying my mind.

I am involved in discussions about creating online tools for local food networks—to help producers coordinate better with distributors/retailers and both with consumers. It crossed my mind that this was a sub-economy, partly functioning on people's decision to buy locally (and organically, etc....) in spite of higher food costs per unit. A lot of the discussions in terms of software involved how orders were made, changed, deadline for weekly orders, whether to pre-sell or charge on delivery, etc.... As well as the possibility of people willingly investing in producer's expansion projects (and maintenance and innovation).

It may have been the recent bitcoin news that spurred it, but I thought that if people who participate in this local food economy were to buy common credits to be used on a common online platform, then the incomes and outlays could be more easily handled and tracked. Like Lonergan's BGF component pitching in money, the economy would grow according to people buying in (obviously they could also cash out too, and this might be a problem when collecting data).

Anyway, I created a diagram analogous to LB's baseball diagram to see if I could envision it in that context.

Having done so, I wondered if you might take a glance and see if anything pops out at you—good or bad. I pasted in my version of the baseball diagram at the top as a reference (for me, not for you!)

Best regards

Lowell

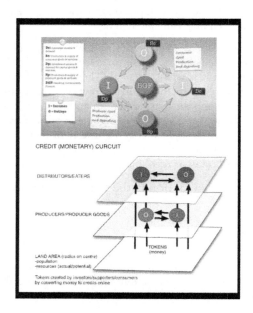

CREDIT (MONETARY) CURCUIT

I answered with a suggestion about re-configuring the diagrams. I continue to get email suggestions about getting in touch with various economists, like Jeffery Sachs and Paul Krugman or contributing to journals. I think at this point in time Lowell's food network is a better bet than Jeremy Sachs or Paul Krugman or another journal article.

3. A Blog Post

In light of the different responses above and others I have not included, it occurred to me that one approach might be to start a blog targeting neither economists nor Lonerganians. So I envisaged a Blog called *Two-Circuit Economics* that I might involve people in food networks and other cooperative efforts might read. My thought is that this increases the pool of potentially interested readers. What follows is a draft for a first entry and is my example of how someone might begin the communication to everyone else, including the food network.

What is the two-circuit approach to economics and finance?

It is not a secret to most that we have a global economic mess and that it has serious implications for the environment and our political stability, not to mention the ability of many to make a decent living. It is also increasingly obvious that economists as a professional group have not figure out why we are in this mess. The evidence is in the continuing existence of competing schools

188

of economics offering contradictory advice. Without an agreed upon understanding about the basic features of production and monetary flows, it is very difficult to come up with policies that truly move towards the goal of reaching a healthy economy that sustains the earth while at the same time providing the materials needed to truly support human development. What we seem to get is temporary fixes that help some (the less than 1%) but hurt many. This is how the 'trickle-down theory' worked. Economists are nonetheless very influential as advisors to governments, banks, think tanks, and corporations. The media seek their advice and the result is that we are all systematically misled about economic realities. You probably know this in your bones. Now it is one thing to be misled about a trivial matter is another to be misled about money. Not that money is the real goal of our existence as the novelist Fay Weldon remarked, "People are as good as their incomes will allow."

This is an extremely complex issue, one for which I do not have ready and easily digestible answers. But I do have a simple proposal for where to start. I believe there is fundamental mistake about how we go about doing economics. The implications of the mistake are quite massive. Mainstream economists, as well as their critics, most assuredly make the mistake. However, if what I am about to propose were to become common coin in our global dealings, I have confidence that there would be clear positive effects for our households and our businesses. And if we positively change how we handle our money and do our business, it cannot but have a positive effect on politics, whether local, regional, national or global. If our economy and politics change, then there is hope for a richer daily life and for the emergence of framework for global concord amongst the people of the world.

Now I am dreaming big. Still, what I think is this: until we sort out the basic mistake taught on the first day of first year economics, even with all the good will in the world, we will not be able to begin to solve in a systematic way the many global problems that beset us. We will be left with piece-meal solutions and ad-hoc strategies. My diagnosis is strategic. There are many causes relevant to the mess we are in, and economics is just one of them. But it is nonetheless a very significant stumbling block. I think it is time to throw some light on the 'dismal science' that is not yet really a science. It is time to point directly at the fundamental mistake in economics and to reflect on its implications. This blog is my effort to do so.

And what is the mistake? Well there are actually many, and as we move along I will address them. But fundamentally we could call the problem a counting error. Macroeconomists systematically treat transactions as if they all more or less same. We have all heard of GDP (Gross Domestic Product). Economists regularly appeal to GDP when they predicting good or bad news. If number goes up, it means we are growing in the right direction; if the number goes does down, things are not going so well. But is simply increasing the total number of sales going to tell us if the increase in actually healthy for us? Are all sales created equally? A common saying in developmental economics is 'give someone a fish and they are fed for the day; teach them to fish and they can feed themselves'. Is the purchase of a net the same as the purchase of a fish? It seems not for the net can catch many fish well past the day of the sale. The fish however must be eaten or it will be no good. That this tells me is that the sale of the net is somehow different than the sale of the fish. The sale of the net should have a different effect on economic development that the sale of the fish. As I hope you will notice for yourselves, there are, in fact, two distinct circuits operating in any economy. One is called the basic circuit: it provides good and services that when purchased enter into our standard of living. The other is called the surplus circuit: it provides the goods that produce the good and services that enter in a standard of living. When we buy a loaf of bread from the baker, the baker transforms so much wheat and other ingredients into the loaf of bread that we consume. Any sales that result are in the basic circuit. However, when the baker buys an oven, the oven becomes the means for many future loaves of bread. These sales are in the surplus circuit. The thing to notice about the surplus circuit is that the materials that become the baker's oven ultimately have a disproportionate effect on the number of relevant transactions. One oven produces many loaves and many loaves are many transactions in an economy. The circuits are related. If the baker purchases an addition oven, he can significantly accelerate the total number of loaves he makes every day. The sale of the oven in the surplus circuit will accelerate production (and hopefully sales) in the basic circuit. You cannot count the sale of the baker's oven the same way you count the sale of the loaf of bread that is its product, and the firm that produces professional baker's oven is not the same as the bakery in terms of its effect on the economy. The difference is basic and verifiable. It is relevant to any business and to any household because they are both linked to both circuits. It is relevant whether we are talking about local, national, or international economic flows, whether we are talking about barter, socialist or capitalist economies, whether we are doing microeconomics or macroeconomics. This fundamental reality of two circuits and two kinds of firm is presently systematically missed by economists and, as long as it is, the figures used to understanding what is going on in our economies will be fundamentally flawed. Appealing to GDP growth as a measure of economic health is about as valid as a physician using blood pressure rates as the main diagnostic tool. Understanding the human body requires much more than knowing blood pressure rates and understanding economies requires much more than using a the crude measure of GDP.

The idea is not mine; Bernard Lonergan, a Canadian from Buckingham Quebec, discovered it seventy years ago. His timing was bad. At that time, the world of economics was absorbing John Maynard Keynes' General Theory and Lonergan could not find a receptive audience for his work among economists. Lonergan's economic writings were finally published until 1995 but his ideas have remained an outlier in the world of professional economics. This does not mean he was wrong, it just means that he has not been read. If you want to read the original source I suggest you pick up his *For a New Political Economy* published by the University of Toronto Press.

What I propose to do in the weeks and months to follow is comment on Lonergan's economics ideas and their implications for understanding economics and economic life. Once we are on firm ground about the two circuits and the two kinds of firms they involve, then we can begin to expand and explore the implications of this idea for other aspects of the economy, such as finance, and to consider how economics is related to the rest of life. We will discover that there is a way to understand economic development that is not necessarily tied to the waste and destruction of natural resources and that in principle it is possible to avoid the recessions and worse that currently plague the normal course of the business cycle. We will also discover that fair distribution and a rising standard of living for all should be the normal, expected outcome of a period of economic expansion. Finally I hope to put to rest the notion that business is about maximizing profit. It is not. Profit has a temporary function, but it is not a permanent feature of a healthy economy. But you will have to stick with me for a while to find out why this is so. I hope you do.

In the next post I will introduce in a simple and easily understood way, what a two-circuit approach to economy is.

The above is probably too wordy and too general for a Blog. And there's no picture. Last year I spoke with a professional Blog advisor who I met on a bus trip into the city. He suggested that a Blog entry needs to be something that can be read in the five minutes. If it's too long it will go unread. That is certainly a shift in context from the academic essay or the classroom lecture and it presents a different kind of communications challenge. James Duffy has suggested ways to re-present and sub-divide what I wrote so that it will appeal to a range of different audiences. I am thinking about it. I hope I can be creative about it. Talking to James certainly helps me imagine different ways to present the material.

IV. Reflection

FS8>C9 (C$_{89}$) communication in economics is a tremendous *creative* challenge.[14] My bus companion suggested I use Twitter but how can you express what needs to be expressed in 141 characters? Do we need a new form of haiku? But then haiku was never the stuff of the daily news and, in any case, the daily news is increasingly a fragmented digital press where each person reads the site which is in accord with their interests and viewpoint in "their divided community, their conflicting actions, and the messy situation…headed for disaster."[15] Lonergan writes:

---

[14] See Bernard Lonergan, "Healing and Creating in History," *A Third Collection*, 100–07.
[15] Bernard *Lonergan Method in Theology* (London: Darton, Longman & Todd, 1972), 358.

It affects the situation, for situations are the cumulative product of previous actions and, when previous actions have been guided by the light and darkness of dialectic, the resulting situation is not some intelligible whole but rather a set of misshapen, poorly proportioned, and incoherent fragments.[16]

Where indeed is the *light* amidst this darkness?

I am led to a simple conclusion. Lonergan was known to quote Newman that "a thousand difficulties do not make a doubt." I have no doubt about the reality of two economic circuits. It is pretty simple and obvious once you've grasped it. I also believe that the *light* is "the image of God" in us. That light is creative, it's an idea, and it's righteous. So I believe with Lonergan in the possibility that

> each stage of the long process is ushered in by a new idea that has to overcome the interested vested in the old idea, that has to seek realization through the risks of enterprise, that can yield its full fruit only when adapted and modified by a thousand strokes of creative imagination.[17]

We simply need to commit to working together in making our strokes and rowing our boats ashore. Our communicated creativity, our communicated imagination is ultimately the source of the solution to the problem, whether it be a haiku or an academic paper, a Blog or a diagram, configuring a food network in light of two circuits, or simple speaking and listening to each other.

And Lonergan continues:

> And every idea, once it has borne its fruit, has to reconcile itself to death. A new idea is only new once it first appears. It comes to man not as a possession forever but only as a transient servant; it has its day glorious or foul; it lives for a period that is short or long according to its generality; but it may be succeeded by other alternatives; and in any case, it will be transformed, perhaps beyond recognition, by higher generalizations.[18]

Lonergan pushed fiercely for a *general* method, a higher viewpoint for collaboration and for economics. As such the birthing has been slow, painful, and as yet incomplete. Yet for all that it may live for a long time even as its expression and operation are transformed beyond what we presently recognize. It is our task to row the boat up to the next bend in the river. We cannot row alone and we do not know exactly what is around the bend. It may be death—well it will be eventually—but we might hope that the death is the seeding of a righteous revolution in human collaboration.

---

[16] *Ibid.*
[17] *CWL* 21, 20.
[18] *Ibid.*

APPENDIX: BETWEEN GORDON AND LOWE[19]

Between Gordon & Lowe

Gordon — very like Keynes plus Hicks

Lowe — complementary to Eaton — stationary state permanent
a) expansion — when prop,
increase because it

Lowe

1. 2 circuits — question on change?
2. 2 production process — related as accelerator & accelerated
3. 2 accelerators — surplus circuit
   = central redistributional (change of ownership &c among)
4. 2 expansion { surplus : "Thrift & enterprise"
                 basic : " ?
6. 2 evasions { = B F T
               Deficit gov't spending
5. 2 exaggerations. { social dividend ignored & profit
                     basic expansion not short
                     because social dividend dropping

2 types of exchange { operative
                     redistributional
2 operative circuits { basic : accelerated
                      surplus : accelerating
2 accelerators surplus circuit : real
               CR : monetary
2 expansion { surplus
             basic
2 precepts { thrift & enterprise
            "missing"
2 evasions { FBFT
            Deficit spending

2 defects
± social dividend = profit
↓ basic expansion
not short because
social dividend
(thought to be profit)
vanishing

[19]Archive File 34440D0E080, available on the Lonergan Archives website: www.bernardlonergan.com.

Michael George

I. Contexts

"Finality, Love, Marriage" was published in *Theological Studies* in 1943. At the time, Lonergan was teaching theology in Montreal, and had taught a course on marriage the previous year.[1] Lonergan hoped to generate some discussion with this article relating to its contribution to the theory of marriage, but in the spring of 1944, the Holy Office "ruled as inadmissible the opinion of recent writers who either denied that the generation and education of children was the primary end of marriage, or taught that secondary ends were not essentially subordinate to the primary."[2] And that, as they say, was that.

There is a basic context of Lonergan the theologian recognizing significant problems in the Catholic understanding of sexuality, attempting to address some of the problems, recognizing all the while that open discourse was highly unlikely to occur. The means by which Lonergan approaches the issue is to present a highly developed and concise framework wherein the relations of the human to the divine (and vice versa), are established in terms of horizontal ends, which are designated essential, and vertical ends, which are designated as more excellent. In fact, there is a nascent metaphysical and historical project sketched out, with accompanying schematic diagrams, that demonstrates the comprehensiveness and coherence of his approach, by indicating how the lower order integrities (horizontal ends) are informed and transformed by the higher orders (vertical ends), culminating in the experience of the beatific vision. The project becomes problematic, from the point of view of the Magisterium, when Lonergan moves from his larger perspective to the actual question of sexuality.

This introduces contexts that situate human sexuality in the reality of history, encompassing evolution, the emergence of humans, the development of social relationships, and their attendant regulations, and germane to the point of the essay, the beginnings of what might become (a) science(s) of sexuality.[3] The discussion of sexuality outside of the religious sphere tends to focus on instrumental concerns,

---

[1] Bernard Lonergan, *Collection*, ed. Fred Crowe and Robert Doran, vol. 4, *Collected Works of Bernard Lonergan* (Toronto: University of Toronto Press, 1988), 258–59 (hereafter *CWL* 4).

[2] *Ibid.*, 263–64.

[3] The simplest image I can think of here is taking McShane's symbolic expression of the person in history, H S+ F (p_i, c_j, b_k, z_l, u_m, r_n) with a singular focus on sexuality. The lack of all but descriptive behavioral models regarding sexuality indicates how far we are from adequately imagining what is required. See Philip McShane, *Wealth of Self and Wealth of Nations: Self-Axis of the Great Ascent* (Washington, D.C.: University Press of America, 1975, 1981), 106–07.

and the attendant observable behaviours. This translates into a concern for instrumentality and for sexual dysfunction, begging the question of a normative understanding of sexuality.[4] Interestingly, and sadly, there is a similar concern with the instrumentality of sexual acts that dominates the Christian/Catholic perspective up until the present time.[5] For students of Lonergan, the recurring issue of philosophies and methodologies that do not, or cannot, provide an adequately explanatory perspective is nothing new. In "Finality, Love, Marriage" (hereafter FLM), Lonergan provides the basis for a religiously and historically comprehensive setting of human sexuality, allowing for an accounting of the facticity of sexual reality and the imminent and potential normative dimensions that impel sexuality to increased meaning and significance, intending ecstatic union. Fr. Crowe points out in his notes to FLM that Lonergan's analysis of love in this article would never be expanded upon, or developed, in his later work.[6]

The strangeness of FLM, understandable[7] enough given the context of theological discussions of sexuality within Catholicism, lies in the disparity between the strength, comprehensiveness and coherence of Lonergan's arguments in the beginning of FLM, and the rather diffuse conclusions that he articulates at its end. The larger argument is situated firmly in the midst of the traditional Catholic teaching, inasmuch as Lonergan concludes that through vertical finality, particularly at the level of reason, the essential end of sexuality, the production of offspring and the raising and educating of adult children, through the formal contract of marriage, also contributes to the development of the parents towards the end of human perfection, through the aspect of friendship, a less essential end of bi-sexual fecundity.[8] Lonergan goes on to make the orthodoxy of his arguments and conclusions explicit: "Now if this analysis satisfies the exigencies of modern data and insights, it is no less true that it leads immediately to the traditional position on the ends of marriage."[9]

However, there are three explicit points in his summation of his arguments that suggest, at the least, unexplored possibilities about the role of sexuality as a primary

---

[4] This is nicely illustrated by the first two books by the acknowledged leaders in the field, Masters and Johnson (whose exploits are now serialized on television). See William Masters and Virginia Johnson, *Human Sexual Response* (Toronto: Bantam Books, 1966) and Masters and Johnson, *Human Sexual Inadequacy* (Toronto: Bantam Books, 1970).
[5] For a contemporary, but less than current, perspective, see *On the Pastoral Care of Homosexual Persons*, October, 1986, by the Congregation for the Doctrine of the Faith, Rome.
[6] *CWL* 4, 261.
[7] "Understandable" here in a weak, and somewhat pejorative, sense.
[8] See FLM, *CWL* 4, 41–7. Note that Lonergan uses the term "bi-sexual" to refer to the male and female in a unitive way.
[9] *Ibid.*, 48.

element in the process toward human perfection. The first point is a lengthy footnote, where Lonergan distinguishes his understanding from that of Dr. Doms' position, whose 1937 book, *Du sens et de la fin du marriage,* provided the initial starting point and rationale for Lonergan to write FLM.[10]

> What is the ontological significance of bisexuality? It is only a terminological difference when he asserts that the meaning of marriage is union and I say that the act and end of bisexuality is union, or when in different ways we both place two ends beyond this union. But when he speaks of this meaning of union as immanent, intrinsic, immediate, I distinguish: in the chronological order of human knowledge or of the development of human appreciation, the union is first; but in the ontological order the ordinations to the ends are more immanent, more intrinsic, more immediate to the union than the union itself. For what is first in the ontological constitution of a thing is not the experiential datum but, on the contrary, what is known in the last and most general act of understanding with regard to it; what is next, is the next most general understanding; etc. Thus the proximate end of bisexuality is union; but of its nature, bisexuality is an instrument of fecundity, so that the end of fecundity is more an end of bisexuality than is union; similarly, bisexual union has a vertical finality to higher unions of friendship and charity; *and these enter more intimately into the significance of bisexuality than does the union on the level of nature.*[11]

The second point follows the footnote quoted above, in the main text:

> If, then, reason incorporates sex as sex is in itself, it will incorporate it as subordinate to its horizontal end, and so marriage will be an incorporation of the horizontal finality of sex much more than of sex itself; nor is this to forget vertical finality, for vertical and horizontal finalities are not alternatives, but the vertical emerges all the more strongly as the horizontal is realized the more fully.[12]

The third point is taken from the beginning of Lonergan's summing up:

> This brings us to our main analytic conclusion. The process of bisexual fecundity (Z, Z', Z'') is in man integrated with the processes of reason and of grace. Such integration takes place by projection, by the incorporation of the lower level of activity within the higher. The incorporation on the level of reason is generically a friendship (Y) and

---

[10] *Ibid.,* 17, note 2.
[11] *Ibid.,* 46, note 73 (emphasis added).
[12] *Ibid.,* 46.

specifically a contractual bond *(Y')*; the latter has a horizontal finality to the procreation and education of children *(Y")*, but the former has a vertical dynamism tending to advance in human perfection *(Q')*.[13]

As the whole argument of FLM is directed toward a re-examination of the Catholic understanding of marriage, the language, perimeters of the argument, and explicit conclusions drawn all assume and affirm the traditional understanding that had existed. From a contemporary and nominal perspective, the absence of any consideration of sex outside the bounds of marriage, homosexuality, bisexuality, transgender sexuality, and the like, might render the relevance of FLM to be limited to a more sophisticated, but nonetheless traditional, rationale of orthodox Catholic teaching. However, it is fairly clear from the points listed above, and from the initial response of the Holy Office, that there was more going on in FLM than a mere justification and reiteration of the traditionally held teaching on marriage. In this context, at the beginning of our attempts to begin thinking and working collaboratively in and with functional specializations, FLM provides a prime example of Philip McShane's call to identify within zones of functional interest, those ideas that are "worth recycling." Beginning to address the issue of sexuality in terms of a normative dimension of growth and development that is not focused solely on instrumental agendas, but rather would recognize and appreciate how sex is a fundamental element in the meaning and appreciation of life as human aspiration would change everything about culture. The empirical evidence suggests that we haven't begun to even consider such a possibility.

A foundational context is the questioning of the suspicion or rejection of the goodness of human desires. McShane suggests that this context, which is expressed randomly throughout the papers presented here, is

> question[ing] the concrete distinction between religious experience and human experience. There is an identification of all human (thus even slimly voluntary) experience as *de facto* religious. What is not such is the non-entity of core aberration. … What is at issue are the rejections of all forms of dualistic thinking regarding human desires' goodness.[14]

The ubiquitous presence of these dualisms is evidenced by Lonergan's conclusion to FLM, where he questions the traditional perspective that privileges virginity over marriage, widowhood over second marriage, and temporary abstinence to use within

---

[13] *Ibid.*, 47.

[14] Philip McShane, electronic communication with the author, November 27, 2013.

marriage.[15] The oddness of this perspective remains an integral part of official Catholic belief, although one might wonder if the (non)practice usually follows.[16]

Other contexts to be considered here would include historical studies, especially those focused on the emergence of social regulations regarding sexuality, patriarchy, the connection between economics and sexuality, and the beginnings of sexual research in the 20th century. It is immediately and generally obvious that there is an almost complete absence of any disciplinary approach that addresses sexuality in a manner that would allow "a normative pattern of recurrent and related operations [that would] yield cumulative and progressive results."[17] I would suggest that FLM provides an initial starting point for considering what such approaches and sciences might look like, and why, in beginning to understand the roles and significance of sexuality. For example, contrasting the importance of communication in human sexuality with the processes of cell mitosis indicates in a minor way the levels of complexity and attendant understandings that we have yet to begin to imagine in order to begin to grasp the evolutionary significance of sexuality, and where it/we might go. Unlike Bachelard, we are not yet ready to begin to build a house.

Following the dictum of GEM 141, I offer the following minimal observations about the personal context of this project. A fair amount of angst and unresolved issues, especially around the Catholic context and tone of the general discussion, incorporating about 30 years of my life, has surfaced. Also, a fair amount of self-reflection, incorporating most of the other 25 years I've experienced, has occurred trying to comprehend my own experiences regarding sexuality. Finally, I am reminded of how difficult it is to operate and function with the intention of working with and within the functional specialties. Collaborative possibilities seem to me mostly vague promises, with moments of lucid possibility occasioned by rare events and opportunities such as this meeting, with kindred spirits providing lift and hope.

II. Functional Operations

Most of the other essays in this volume illustrate particular specialties and the handing on. This essay may seem to be exceptional yet it is not. Nor does it belong, like some of the broader essays, in the zone named by McShane as C_9. The effort here is foundational and, as we shall see in the next section, its handing on can be considered as C_{5i}, where the subscript i there ranges from 1 to 8, and even to 9. That communicative effort, one of accelerating the cycle in particular zones, is one of the

---

[15] *CWL* 4, 49.

[16] For an official set of statements, see Articles 1618–1620, the section entitled "Virginity for the Sake of the Kingdom," in the middle of the larger section dealing with marriage, in the *Catechism of the Catholic Church, Second Edition* (Washington, D.C.: U.S. Conference of Catholic Bishops, Libreria Editrice Vaticana, 2000).

[17] *Method in Theology* (London: Darton, Longman & Todd, 1972), 5.

functions of foundations. The other *per se* function of foundations is fantasy, the stretching of the foundational person's molecularity, especially, obviously, the neuropatterns, into realms of human liberation and destiny. It is this second function that is the driving force behind the foundational effort. A reach to what might be, and might be good, one that includes glimpsing reality "better than it was."[18] So, this section might be considered to focus on the component of the matrix of communication designated by C_{55}. I ask, then, for foundational persons to rise up from the present best standard model to a lift into pilgrim perspective that echoes earthily the stand expressed in the title *The Everlasting Joy of Being Human*.[19]

The project, then, is about starting to communicate about fundamental readjustments that are required if life is to become normatively human. Some twenty plus years ago, I recall having a discussion with a friend about normative points of reference, as social and cultural indicators, and he, as a good student of Lonergan, reminded me that given the longer cycle of decline, even good and discerning people had great difficulty in discerning the bits that still made sense, and that given current trends, things were only going to get harder.[20] The fantasizing, then, becomes the starting point of the possibility of reorientation in a constructive way, but the motive for such imagining and projecting lies in a desire that is sufficiently strong to create within us an unease with our disease. The topical sentence that highlights this in FLM is, "The ignorance and frailty of fallen man tend to center an infinite craving on a finite object or release: that may be wealth, or fame, or power, but most commonly it is sex."[21] Why sex? Because "(o)ur being is a loneliness in search of a mediation of loneliness toward an ultimate transformation of loneliness."[22] And connecting with people, however dysfunctionally, is still a more adequate response to that craving than the other options listed, partly, I believe, because we recognize that infinite craving in others. McShane goes on in *Wealth of Self* to quote Hermann Hesse's *Steppenwolf*, where Hermine "is trying to explain to the forty-year-old Harry Haller why his life is nothing, and yet not nothing."[23] The aspect of craving that is fundamental, and personal, and unrelenting requires some level of self-recognition,

---

[18] *Method in Theology*, 251

[19] Philip McShane, Axial Publishing, 2013. It has taken me months to start to make this part of myself, but I think that's what is required. Part of my reckoning about GEM 141, this is the part of the personal context that has felt like growth, albeit haltingly. The simple title initially, and indeed continuingly, reads to me like a manifesto, one "worth a life." I trust this, and couldn't have arrived at it myself. Philip McShane is the conductor here, and all here/hear know why.

[20] The friend was John Dool, who shortly after this discussion, took up his current position at St. Peter's Seminary, in London, Ontario.

[21] "Finality, Love, Marriage," *CWL* 4, 49.

[22] *Wealth of Self and Wealth of Nations,* 110.

[23] *Ibid.*, 110.

and critical appraisal if it is to lead to anything other than more or less efficient means of self-destruction. At least two of the papers presented to this conference address *craving*, one within the theological/philosophical realm, and the other within the project of neuroscience.[24]

The ignorance and frailty that we are craves sex; and sex is finite, as are we. But our craving intimates our longing for the infinite, and a sexuality that would intend human perfection, and beyond, need not be considered an obstacle, but rather an opportunity for growth. The list of nine separate sets of considerations that should inform the transcendentally functional subject who is attending, inquiring, reflecting, planning and deliberating—and sexual all the while—may provide some fuel for fantasy in projecting foundational beginnings of a potentially normative sexuality.[25]

The following suggestions could initiate any number of discussions, options or alternatives, but it seems that the more deliberate the engagement, the greater the potential for creativity, which is turn shifts probabilities, creates new conditions to consider, and so on.

(1) Each of the different kinds of conscious operations that occur—attending to sex, noticing urges, stimuli, scents, tactile sensations, visual cues, non-verbal communication, etc..., and likewise, for inquiring (sexual curiosity, response, etc...), reflecting (which should include heightened appreciation, greater intimacy, etc...), planning and deliberating (let's do it again, but differently, slower, outside, etc...).

(2) the biological, aesthetic, intellectual, dramatic, practical, or worshipful patterns of experience, where sex is reconsidered, or initially addressed, in terms of expanded possibilities wherein focus is directed, and moved from one pattern to another, and one set of operations to another. While our appreciation and skill with neuro-heuristics is limited, it would seem that sex would provide a good way to develop these skills and that the rewards would be, well, worth the effort.

The rest of the seven differentiated headings would provide interesting grist for reconfiguring sexual assumptions, practices, and assessments. Additionally, most of this work would be more adequately undertaken in a collaborative manner. This is the quintessential win-win scenario.

---

[24] See, respectively, James Duffy, "A Special Relation," and Bob Henman, "Functional Research in Neuroscience."
[25] See the nine sets which are provided in order to differentiate the basic terms and relations in the "Foundations" chapter, under the heading of 'General Theological Categories.' *Method in Theology*, 286–287.

The project is existential in the deepest sense. It pits persons against the accumulated distortion, fear, shame of our historical heritage that we appropriate in our socialization and various educations. As McShane puts it,

> [M]y aim is to point discriminatingly to work in the full cycle. It is the massive task of rescuing humanity's sexual nature, so that sexuality in itself may blossom as the seed and **show** of infinite craving and that, in relation to the other bents of ignorant and frail humanity, it frees humanity to focus on the brutalization of desire in its finding a narrow home in wealth, fame and power.[26]

The current impossibility of such a focus informing the study and normative appropriation of sexuality (skills, mastery, etc.) is even more problematic when situated within the realm of theological discourse and reflection. Quite apart from the lack of adequate intellectual models within the natural and social sciences, the project is immediately at odds with many of the existing (anti)foundational perspectives that inform religious sensibilities. Further, the obvious commodification of sexuality in the larger culture is in the largest part made possible by the reductionist and instrumental categorization of sexuality first generated, and then maintained, by religion. Foucault's old dictum, "who benefits from telling this story?" reverberates loudly.

III. The Handing-on Process

And so, from a poor bastardized version of C_{55} to C_{5i}, where we start with C_{56} and try to envisage how to accelerate the cycle in the zone of doctrines. First, God help us. It seems to me, that if a small group of foundational people were to pick a strategic starting point to shift the probabilities toward doctrinal readjustment, that a good starting point from within theology would be a public examination of the papal encyclical *Humanae Vitae*. This encyclical, published in 1968, is generally considered to be one of the weakest arguments presented defending the primacy of procreation and especially the proscribing of the use of birth control within marriage (the only recognized legitimate way to have sex). The encyclical is even more famous (amongst Catholics, especially in North America and Europe), as being ignored by more practicing Catholics than any other previous encyclical. This is as clear an indication as can be had that, at least on matters pertaining to sexuality, the Magisterium and the laity live in different worlds. If the application of the theory (what is and isn't permitted during sex) isn't recognized as legitimate, or authentic, even with the inadequate framework of understanding sexuality we have at present, then the legitimacy of your doctrines, and by extension, your foundations, are equally called into public question. It seems to me that given the lack of adequately scientific

---

[26] Philip McShane, electronic communication to the author, November 27, 2013.

disciplines to provide "cumulative and progressive results" that identifying the nature of the existing problems clearly and comprehensively (the what questions), is the necessary first stage, before any normative claims and policies can be situated and implemented (is, or what to do questions and projects). This has the effect of highlighting the primacy of the first four functional specialties, which, in turn, suggests and demonstrates the essentially historical origins of our collective sexual inadequacies.

The discussion around implementation, represented by a mix of C_{56} and C_{89} discourses, where everyday policies are formulated and implemented would be most profitably focused on the obvious lack of policies with substantial positive content. Apart from the exhortation to intend to procreate every time you have sex with your lawfully and sacramentally bound spouse, there aren't a lot of positive alternatives or options, either suggested or intimated. The primary policy, even within Catholic marriage,[27] is "don't," which doesn't provide a whole lot of anything with which to build positive, user-friendly, sexually positive and affirming policies. Not surprisingly, in American states where conservative Christians have a significant political influence, sexual education also seems to adopt the minimalist approach, and "don't" seems to cover most of the curriculum. Substantive discussions around content and pedagogical approaches to sexuality would seem to provide good starting points for the redressing of policies.

Communication, the C_{58} zone, is the most problematic zone in an immediate fashion, inasmuch as the sexual biases that plague us are part of our linguistic sensibilities and structures. Again, and only partially hypothetically, pedagogical approaches to child raising that introduced positive sexual narratives (content and substance still very much to be determined) and communication wherein parents and educators (inter- and extra-institutional, formal and informal) became self-censors regarding all expressions related to sexuality (eliminating as much negativity as possible, and providing/creating life affirming images and metaphors to promote sexual appreciation) would be enormously important. The immediate impact on consumer advertising and most of what passes for main-stream entertainment that such developments would have is hardly imaginable. Other immediate concerns would have to encompass the lack of a positive set of terms, images, or metaphors that would shift neuro-dynamic processes in a positive manner. Perhaps a C_{58} development of sorts, but I have an even harder time imagining what sexually positive terms and images would consist of. The lack of such terms and images hinders the possibilities of constructive exchanges between the functional specialties (even potentially) in ways that can only be described as debilitating.

C_{59} is pretty much where we live right now. I may be overly optimistic in terms of this zone, but I believe that the growing number of Gender and Women Studies

---

[27] See above, n. 16.

programs, even in these days of increased corporate control of educational institutions, is a sign of hope. As a graduate student in theology, I recall my Feminist Theology course as being the most active and engaged class that I had encountered. It may also have been partially due to the fact that the course was taught in French, and that the majority of students were Francophones. The different rhythms of language affect the patterns of our sensibilities differently. The recent World Cup matches in Brazil demonstrate this in a tangible fashion. Which country plays "the beautiful game?" This time, at least, it wasn't Brazil, which came as somewhat of a surprise to me. Costa Rica? Who knew? The existence of programs, teams, projects, which systematically challenge the hegemony of the status quo has inherent value in generating dialogues that systematically seek alternative accounts, narratives, and rationales. Hell, some people even try to develop a control of meaning. Who (k)new?

Back to the beginning, C_{51}, where research might profitably start by noticing the data that isn't readily available, or even considered. The discussion above, in C_{59}, indicates this as a possible strategic move in an attempt to recalibrate the beginnings of the "what is actually going on" question that deals with the factuality of a debased cultural grasp of sexuality. Foucault's "who is benefitting from this (dominant) story?" again seems relevant. Another relatively straightforward starting point for this type of undertaking might be to start to collect and organize the data that clearly obtain between the ostensible rationales that shape society (government, economics, education, social and health services, religions, etc.) and the implementation and lived experience that results from such agendas. Maybe Virginia needs to reassess Santa Claus, actual or imagined? Research that identifies existing disparities, in all dimensions of life, becomes critically effective at raising the better sorts of questions.

The people with (potentially) good ideas, the C_{52} interpreters, don't have an abundance of constructive resources to work with in the field of sexuality. "Don't" as the current expression of sexual agency doesn't seem to promise much in the way of creative possibilities. Given the existing climate, I wonder if FS_2 people might not have to have an equally well-honed skill set in one or more of the other functional specialties. I suppose that in periods of extended historical decline that this would be true for most of the specialties, notwithstanding the lack of an appreciation of the necessity for collaborative cultural relations that holds generally. As noted earlier, contemporary research into sexuality tends toward the instrumental and behaviorist models. As these reductionist frameworks tend to discourage "thinking outside the box" (or thinking at all), it is difficult to project genuinely imaginative responses arising with any serious chance of adequate comprehension and adaptation. As a group most suited to function under these conditions, I would think that the better interpreters were artists, dancers, musicians, and the like. Those persons with aesthetic skills and perceptions are the most likely group to be able to shape novel responses to questions when our lives tend to be lived under increasingly stressful conditions. However, in a similar condition to Lonergan (I imagine), I recall the somewhat apocryphal story told of John Cage, who was asked by an interviewer if his lack of popular appeal and a sizeable audience bothered him, and he responded by saying that he wasn't much bothered, primarily because his audience didn't exist yet.

The historians meshed in C_{53} have the responsibility of crafting more coherent narratives, within which human existence can become (hopefully) more livable. The requirement to give shape to human experience in a comprehensive and comprehensible fashion demands as much creativity as can be mustered. I find the exhortation to view the past as "better than it was"[28] a fabulously outrageous idea, but a necessary one given the centrality of history for consciousness. Perhaps with specific reference to sexuality, the larger text of evolutionary development of life and its dissemination to popular and specialized audiences might permit greater and more receptive appreciation of the evolutionary capacity still to be identified and actualized in sexuality. Also, given such an evolutionary narrative, it would seem to provide a conducive and appropriate venue to introduce recurrent schemes and emergent probability in such a way that the increased control of meaning becomes an obvious stage in human development. Sex that connected persons to their own selves and others in ways that enhanced all the constructive possibilities that are available, actually and potentially, would be a history worth working for.

The dialecticians are the fix-it people. In C_{54} you have to have the most self-actualized, authentic sorts available, or it all goes to hell without a handcart (in a hurry). For sexuality, the control of meaning required means that some poor soul has to be able to quickly and effectively master the ranges of possibility that Lonergan outlines in pages 286–287 of *Method in Theology*, basically requiring the full skill sets required to function in all the specialties, and not be overwhelmed by the systemic barbarism and less than human conditions that characterize most human experience and relations. These are not dime-a-dozen types of people. For those relatively intact, aware and functional, the counter-positions for sexuality are relatively easy to identify and diagnose. Reversing them is the hard part, and the small and woefully inadequate suggestions raised in this paper aren't sufficient to even rate as inadequate. Woefully, potentially, inadequate suggestions? Something like that. Or worse. These persons have the task of orienting the historical project in a hopeful and constructive way. Unfortunately, in the case of sexuality, there are no sufficient comprehensive positions to be generally endorsed that exist at present. The dearth of positive references is mirrored, not surprisingly, in the generic field of human health, where the constructive model of a healthy human being primarily consists of the absence of identifiable ailments and illnesses. In sexuality, reversing the counter-positions can start with a straightforward response. Don't "don't."[29]

---

[28] *Method in Theology,* 251.
[29] I felt this was simpler (more user friendly?) than my revision of Nancy Reagan's motto, "Don't say no!"

IV. Concluding Reflections

I think that the most interesting thing that I have become aware of in the process of putting this paper together is the conspicuous absence of people asking the right kinds of questions. Hand in hand with that, is my recognition of how much easier it is to critique than to construct, which reaffirms the significance and inherent difficulty of the two dimensions of initial foundational possibilities, the attempt to fantasize and the communicative effort of accelerating the cycle in particular zones. These zones don't actually exist yet, throwing us back to fantasy.

Another point of interest is that I found it humorous, repeatedly, recognizing the number of inadvertent double entendres in Lonergan's text that had sexual connotations and that probably wouldn't have occurred to Lonergan when he published FLM in 1943. I started to see/find the same types of inadvertent (?) slips in my own notes and wondered if the slips were occurring because I was thinking about sexuality, or that much sexual language/discourse consists of slang and metaphorical sleight of hand because of the reluctance/inability of our culture (most cultures?) to openly identify explicit sexual themes, topics and issues. The continual and persistent preoccupation of humans with sex and sexual urges seems to be generic and ubiquitous. However, in the 1990's, I recall a text that was based on extensive interviews about sexual beliefs and behaviours in the United States, and the book's primary conclusion, based on their statistical evaluation of the data that they gathered, suggested that most Americans did not have regular sex (on average, less than twice a week), and worried about it due to the widely held belief that other Americans were having a great deal of sex, which fueled their own insecurity about their perceived lack of sexual opportunities. This perception, however, did not alter the low statistical amount of actual sexual activity. All of which led me to assume that some sort of American (maybe more societies?) were suffering from some type of sexual *ressentiment*. In part, the respondents identified the extensive exposure to sexual innuendo and agendas in popular media as fueling their insecurities. But, they believed that their neighbours, friends, etc. were all enjoying more, and better, sexual relationships/activities than they themselves were. Again, instrumental and behavioral agendas/models formed the basis of the study, and the type of data that was collected. Very strange, indeed.

Another anecdote occurred to me whilst reading and thinking about this material. In the 1990's, I taught on a fairly regular basis an undergraduate course entitled "Religion and Sexuality." When discussing with the class some behavioral statistics around types and frequencies of various sexual activities, I took an impromptu poll of the class (of around 35–40 students) about generic sexual preferences. I was stupefied when more than three quarters of the class identified their sexual fantasies as more rewarding/fulfilling than actual sexual activities with a partner. I am still shocked by that response, and I recall wondering what conceivable roots and origins could generate such pathological inclinations. What was even more surprising was that the class could not understand why I thought their responses were so outrageous. They tried to convince me, with a number of arguments, that their responses favouring sexual fantasies were more reasonable than preferring sex with another human being. Humour can only take one so far, and I remain flabbergasted to this day.

In my attempts to imagine the reality of sexuality in its multi-layered complexity despite the absence of adequately comprehensive scientific disciplines, I attempted to work out some parallels using Lonergan's chart on the human good, found on page 48 in *Method in Theology*. In an overly simplistic fashion, I assumed that if vital needs are primary and essential, and at a minimum, one could assume that food, shelter, and sex were fundamental human requirements, it would be interesting to reflect on the structural complexity and overt cultural identification of the place and significance of the respective vital needs. I was struck by the disparity between food and shelter, with their attendant skills, professions, cultural legacies, literatures, academic and professional studies, etc. and the comparative lack of cultural evidence to support any type of developmental or constructive model/agenda/project of sexuality. Historically, when comparing the evidence for human development and growth of sophistication in the areas of food and shelter there is a great deal of evidence that is lacking when one considers sexuality from a developmental or culturally informed model. I find this to be symptomatic of some of the larger problems talked about earlier in the paper, and also indicative of the lack of historical resources with which the problems might be addressed.

As a conclusion of sorts, I am fairly well convinced that creative and constructive developments in recognizing, appreciating and appropriating sexuality are unlikely to take place initially within the sphere of religion. Given the ungrounded nature of most theological/religious discourse, and its apparent inability to take such fundamental human motives such as craving seriously as incarnate and valuable, it is difficult to imagine that a major reversal in perspective is possible. Without identifying and rejecting the fundamental dualisms that inform at least the origins of Christian thinking and reflection, reversing the existing counter-positions cannot take place. Conversely, and even more problematic, is the realization that some sort of transcendental perspective is required if sexuality is to become anything more than a generally functioning, more or less adequately socially acceptable set of biological urges and reproductive capacities. As I have experienced the difficulty of discussing or considering the normative dimension of sexuality when the optimal model presented and valued is based on virginity (Don't "don't"), the resulting possibilities are stunted and distorted. It is clear to all who give sexuality any prolonged consideration that meaning, growth, and intimacy necessarily require more than biological responses, and that until this is recognized our lives will remain pale shadows of what we might otherwise become: gloriously, ecstatic and joyful aspirations.

Meghan Allerton

I. Context

> There is little doubt that humans are now a major evolutionary force on
> Earth, and that our activities dominate its ecosystems, both on land and
> in the sea.[1]

The global impacts of our activities have got ecologists thinking globally: how are
we altering planetary-scale systems and processes, such as climate and
biogeochemical cycling? How are ecosystems responding? What are the implications
for the future diversity of life on Earth, including human life? And how can we live
sustainably? More and more of the work of ecology is going towards answering such
questions. Here I point briefly to some aspects of that work that provide the context
for my own effort in this essay.

Technologies like NASA's Earth Observing System satellites allow us to
monitor global-scale environmental change, providing data on climate, surface
temperature, land cover, net primary productivity, ocean circulation, atmospheric
chemistry, etc.[2] Observed trends are the basis for further research and data
collection going on around the world in an effort to understand the causes of global
change, and the consequences for organisms and ecosystems. Computer models are
used to project future environments, simulate ecological processes, and estimate the
relative importance of the drivers of change. Models of global biodiversity tell us
that climate change, land use change, nutrient loading, invasions by exotic species,
and atmospheric CO_2 enrichment (all of which are caused by human activity) will be
major drivers of ecosystem change during this century.[3]

There are ongoing efforts to synthesize and assess the findings of ecological
research, monitoring, and modeling. International bodies like the Food and
Agriculture Organization, UN-Water, and the Intergovernmental Panel on Climate
Change (IPCC) produce regular reports on the status of scientific understanding in
different areas of ecology. Another UN initiative, the Millennium Ecosystem
Assessment (MEA), aimed to assess global ecosystem change and the implications
for human well-being. The report warns that we are degrading most of the essential

---

[1] William H. Schlesinger, "Global Change Ecology," *TRENDS in Ecology and Evolution* 21,
no. 6 (2006), 349.

[2] See http://http://eospso.gsfc.nasa.gov.

[3] See Osvaldo E. Sala et al., "Global Biodiversity Scenarios for the Year 2100," *Science* 287
(2000), 1770–74.

services provided by ecosystems (e.g., clean fresh water, local climate regulation, pest control).[4]

This work also sheds light on some problems in the discipline. The MEA report notes that many questions about ecosystem services could not be answered due to "major gaps in global and national monitoring systems," "major gaps in information on nonmarketed ecosystem services," the lack of a "complete inventory of species," and "limited information on the actual distributions of many important plant and animal species."[5] Information gaps mean that there are areas of ecology, especially marine ecology, that are very poorly understood. And even where new monitoring tools leave no shortage of data, there is the further difficulty that

> to be of use to science, the data must be correlated, calibrated, synchronized, and updated. *Wired* observed that 'Earth is peppered with high-tech monitoring hardware, from polar-orbiting satellites to instrument-laden buoys. Problem is, they're all operating in Babel-style disconnect.' Efforts are under way to link everything in a mutually intelligent way via a Global Earth Observation System of Systems.[6]

As it is, these problems hinder efforts like the MEA and IPCC reports in their ability to communicate relevant information to policymakers and answer policy-relevant questions about reversing, mitigating, or adapting to ecosystem change. Environmental policies are created based on the information that is available, or, quite often, by deliberately ignoring it. However, when many parties are involved in reaching international policy agreements, reliable and consistent reports are essential to the creation of policies that effectively address environmental problems.

There have been some successes with international environmental policy, perhaps most notably the 1987 Montreal Protocol to protect the ozone layer. Signed just 18 months after the discovery of the 'ozone hole' and ratified by every member of the United Nations, it has achieved a 98% reduction in the use of ozone depleting substances around the world.[7]

But successes like this are exceptional. Efforts to reach international agreement to mitigate climate change have been largely ineffective. The United Nations Framework Convention on Climate Change requires countries to reduce greenhouse gas emission to "prevent dangerous anthropogenic interference with the climate

---

[4] Millennium Ecosystem Assessment, *Ecosystems and Human Well-being: Synthesis* (Washington, DC: Island Press, 2005), 1.

[5] *Ecosystems and Human Well-being: Synthesis*, 101.

[6] Stewart Brand, *Whole Earth Discipline: An Ecopragmatist Manifesto* (London: Atlantic Books, 2010), 279.

[7] "Montreal Protocol—Achievements to Date and Challenges Ahead," United Nations Environment Programme, http://www.undp.org/content/undp/en/home/ourwork/ sustainable- development/natural-capital-and-the-environment/montreal-protocol.html.

system,"[8] but sets no legally binding limits to emissions. The later Kyoto Protocol was an attempt to set binding emission targets for developed countries only, but the United States did not ratify, Canada has withdrawn, and several other countries, including Russia and Japan, are no longer participating. As a result, the top greenhouse gas emitting countries are without binding targets, and global emissions continue to accelerate.

This is just one of many examples of the failure of policymakers to pick up on the cues coming from research and reports in ecology, though it may be a critical example for the future of Earth's ecosystems and humanity. With climate change the stakes are high. We are already seeing more frequent extreme weather events, increasing drought and desertification, significant losses of biodiversity, food and water shortages, and migration from areas inundated due to rising sea levels. The IPCC now recognizes these impacts as factors that exacerbate existing security issues and increase the risk of conflict around the world.[9] We can expect that these problems will only become more severe in the coming decades, and that the capacity of the Earth to sustain our current population and patterns of consumption will be greatly reduced.

The gravity of the situation and the lack of action to effectively address it are compelling more and more ecologists to speak out, warning us about the dangers of continuing down the business-as-usual track. But their voices have to compete with the anti-science, environment vs. economy, propaganda from mega-industries that profit by keeping us in the dark. Still, those most directly affected by environmental degradation, usually marginalized, low-income, and often Aboriginal communities, are being joined by a growing number of people concerned about climate change and other environmental issues to protest government inaction and corporate control of government policy making.

Around the world there are growing movements. For example, 350.org calls for action on climate change through "online campaigns, grassroots organizing, and mass public actions … coordinated by a global network active in over 188 countries."[10] I have found that the work of 350.org can inspire, educate, and connect people, yet even within local groups there is major disagreement over the solutions. Geo-engineering? Nuclear power? Solar? Wind? Organic food? GMOs? The debate is heated.

How are we to move forward? There is a clear need for a more adequate response to the ecological situation: "Whether it's called managing the commons, natural-infrastructure maintenance, tending the wild, niche construction, ecosystem

---

[8] United Nations, *United Nations Framework Convention on Climate Change*, 1992, Article 2, available from http://unfccc.int/resource/docs/convkp/conveng.pdf.

[9] IPCC Fifth Assessment Working Group II Report, presented at the press conference held in Yokohama, Japan, 31 March 2014.

[10] "What We Do," 350.org, accessed April 2, 2016, http://350.org/about/what-we-do.

engineering, mega-gardening, or intentional Gaia, humanity is now stuck with a planet stewardship role."[11]

II. A Key Message

There is an advance called for, a unified global effort aspired to by Stewart Brand and expressed eloquently in the final quotation of the first section. The need for unified action in response to global environmental problems is easily recognized (there is talk, for example, of climate change as the greatest "collective action" problem faced by humanity), but the full magnitude of the needed shift in culture is not easily recognized, nor can we push towards features of that recognition here. I focus instead on a key insight regarding the needed advance in ecology, but in a way that may help toward that better recognition in a span of inquiries.

Arne Naess was on to facets of the solution in his identification of four "levels of deep ecology."[12] A doubling of Naess' levels, however, brings us closer to the full structure of collaboration that was discovered by Lonergan: a division of labor into eight tasks, four past-oriented, four future-oriented, and not just a group of tasks but an ordered sequence of tasks. Lonergan's breakthrough was to recognize these tasks as "functional specialties, as distinct and separate stages in a single process from data to ultimate results."[13] His breakthrough recognition of the need for a collaborative structure was achieved through his reflections on the state of theology as one of the "academic disciplines" that he eventually wrote of at the beginning of *Method in Theology* as needing to face the future in a fresh creative way. The key insight here is that ecology must face the same challenge. But facing it, or even thinking of how it might face it, has the possibility of throwing light on the challenge of theology and so of enlarging the perspective conveyed by Lonergan's treatment of the issue in *Method in Theology*. There is a sense in which we can notice that broadly and immediately by thinking of theology as having less visible effects than ecology, being, as it were, an ecology of the spirit, an ecology of the essentially invisible.

First, then, we note that the ecological movement has, from the beginning, taken a pragmatic turn. The question is one of a turn-around from destructiveness of the earth and its resources. It does not fit, therefore, with the image of a detached scientific inquiry, yet no one doubts but that it has the bent of a science. Moreover, whether one thinks of Linneus or Darwin, one can see the shadow of that pragmatic bent. Both classification and origin have the shadow of pragmatic interest on them, a shadow that comes from the early ventures in agriculture or animal husbandry. So, one can easily sense the presence of the question, 'what might this be?' right across

---

[11] *Whole Earth Discipline*, 275.
[12] Arne Naess, "Deep Ecology and Ultimate Premises," *The Ecologist* 18, no. 4/5 (1988), 130.
[13] Bernard Lonergan, *Method in Theology* (New York: Seabury, 1979), 136.

all areas of ecology, and when we turn to dividing the labors of ecologists functionally there is no oddity in finding that presence from the beginning. When, then, we turn to the first four paragraphs of chapter one of *Method in Theology* with ecology in mind, it leads us to a better grip of the full challenge that Lonergan seeks to identify when he writes of the need for a new "third way"[14] of inquiry. There are aspects of ecological inquiry that fit in with the notion of successful science mentioned on *Method* page 3, yet so much of present ecological reflection is in the world of "academic disciplines," of discussions around a range of opinions about what went on, what is going on, and what should be going on.

This freshens the perspective on research that Lonergan introduces in *Method*. Research in ecology is a task of obtaining relevant data, including written data, observations and measurements made in the field or laboratory, data from monitoring technologies like NASA's EOS satellites, computer models of ecosystem processes, climate etc. But, more than other fields, the research is directed, seeking a road forward. Furthermore, it is a research that demands an up-to-dateness in all the related fields of inquiry. So it is not difficult to envisage the need for a standard model, one that reaches out heuristically into other domains, indeed one might claim into all domains, from the physics of winds to the follies of human behavior. Further, there is the aspect of feedback, altogether more evident than feedback in theology, due to the annual rhythms of the earth. Last year's crop, the fruit of perhaps a cycle of collaboration, flow thematically into this year's research.

I must be brief in this short essay but even such sketching can ground optimism regarding a better integral view coming from ecological need that yet throws light on the general need, whether it is in theology or in physics. So, when we turn to the next possible stage of collaboration in ecology, that called generally interpretation, there are suggestions in the need for ecological interpretation that widens our view of interpretation. Indeed, a full book-length view of ecological interpretation would go a long way towards fulfilling the aspirations expressed in the first footnote of chapter 7 of *Method in Theology*. Ecological interpretation requires a full context that would sublate studies of origins both of species and of species of opinions and so it would push us back from the simple presentation of the seventh chapter of *Method* to a complex sublation of the third section of chapter 17 of *Insight*. There is already a decent descriptive history of misinterpretations and good interpretations of uses of fire, air, earth and water.

Ecological functional history would aim at a grasp of the story of the application of such interpretations, and it seems to me that such a telling helps to make clear the distinction between, on the one hand, interpretations and their history and, on the other, applied interpretations and their history. So, for instance, the story of the destructive ideas made operative by multinational corporations is brutally

---

[14] *Method in Theology*, 4.

concrete and already available in elementary suggestive forms. Furthermore, that story, or any ecological story, makes pretty evident that the climb through the specialties involves a convergence of interests of different disciplines. The automotive potentials of physics cut into the patterns of the meadows and musics of humanity.

But it is when we face the task of dialectic that the full concreteness emerges and has to be faced. The facing of this difficulty, I suspect, will be less difficult in the ethos of present ecology than in other areas like literature or economics. What does *Assembly*—the first stage of dialectic—mean for an ecologist? The present mood is that we really cannot exclude anything. Dialectic has to assess ice ages and oil transportations, plant mutations and river dammings, and it must assess the patterns of decisions that weave round such realities. But 'it' is personal: it is an ecological sub-community that is sufficient in elderhood to share this global task. This sub-community, in the concrete process of its emergence and operation, will be self-selecting. Present debate—embryonic dialectic at best—offers a large community of interest that will cut itself back to effectiveness in so far as it genuinely tackles the task. That cut-back process would make concrete the proposals of McShane's *Futurology Express*, chapters 8, 9, and 10, and could well take the lead in helping us forward to envisage more broadly the reality of that effort.

When we come to think of foundations, we find a seeming identity of interest in stating the grounds of the future in present popular concern and in the project of functional foundations. But the seeming disappears when we reflect on the naming of the goal in the two patterns of concern. Popular concern talks, in commonsense overreach, of a goal of sane global order. Functional foundations is to be a constant push for detailed improvement in a metascientific heuristics of that order and the push for it. There is no difficulty in naming the foundational effort: it is named in the lists Lonergan gives in *Method in Theology*, pages 286–87. In that list there is metaphysics, but the metaphysics is now enlarged to include a heuristics of functional collaboration. That named heuristics points to a tremendous climb for the global academic community generally, and we turn, in the next section, to the problem of beginning that climb.

But to get a glimpse of the problem and the climb, I would like us to shift our attention from the fifth specialty of Lonergan to the eighth specialty as common sense, so to speak, calls to it in its descriptive foundations. That call lurks in our first section and when it is spelled out locally—think of the mention of situation six times on page 358 of *Method in Theology*—it brings out the shocking complexity of the perspective needed. One might say that the ecological problem varies acre by acre round the entire globe of land and water, and indeed ranges above those acres beyond the sun. Bernard Lonergan wrote eloquently about the break from classical culture, from some ideal of "one solution fits all." Here we are nudged, very concretely, towards the fullness of that break. Ecology gives us a much more easily reached view of the complexity of communications that hides behind Lonergan's apparently simple treatment of it in chapters five and fourteen of *Method in Theology*. And thinking that out—we turn to that in the next section—will backfire into the previous specialties.

So, we think of Communications as selecting creatively from ranges of possibilities, but now we find that uniqueness of local conditions pushing back to enrich the envisaging of possibilities that is the task of systematics. This lifts us far from the old view of system, especially that which is associated with systematic theology. The new view of system has, of course, been treated by others here, following sketchings by McShane. We see especially from ecology the relevance of a geohistorical systematics that is richly diagrammed. Patterns of ancient rice-cultivation or medieval field rotation can be found to be freshly significant in the move away from the cultural interferences associated with mass production.

What, finally, of doctrines? Here we find the same problem as we noted in foundations. At present it is difficult to distinguish doctrines as they are thought out in the sixth specialty and doctrines or policies as they are commonly thought of. That common meaning is caught in the manner in which we can think of the parallel between the last three specialties and the demand of the three directives of decision-making: policy, planning, and executive reflection. In this early stage of seeding the work of collaboration, it seems plausible that we have to tolerate a confusion of the two types of doctrines. We cannot identify the doctrines and metadoctrines that are grounded in foundations in a clear and operative way here and now. That is a thing of the future, as it concerns the stability of a standard metascientific model.

III. Handing-on the Advance

The challenge here, following the general formula for the contributions to this volume, is the challenge of handing-on. In the present essay, that handing-on is not from one specialty to the next but a commonsense problem of commonsense handing-on. The previous section points to aspects of ecology that might make it fertile ground for our efforts to seed future effort. There is, at least, a sense that "something must be done" to stop the destructiveness of the Earth and its resources, and furthermore, a growing realization that the solution will be more complex than reducing the concentration of CO_2 in the atmosphere, stopping clear-cutting of rainforests, or over-harvesting of the oceans.

I think of a recent interview with Bill Moyers, in which renowned environmental activist and science broadcaster David Suzuki reflected on the failure of environmentalists to bring about a needed change in culture. Despite "great victories in the 1970s and 80s, stopping clear cutting of these forests or drilling for oil off the coast of British Columbia ... here we are 30, 35 years later and we're fighting the same battles ... we've failed to shift the perceptual lenses through which we see our place on the planet."[15] The failure points to the need for a new way of meeting the ecological situation, a "third way"[16] that would bring about the broader and more precise shift of our second section. How might that practical division of labor be handed-on in ecology?

I have been helped in my thinking on this question by my fellow authors, and Bill Zanardi homes in on the central problem here: "How do we persuade others this division of labor is worth a try?"[17] Like Zanardi, I am reaching for "concrete strategies for implementing functional specialization well before it is accepted practice,"[18] but I focus on the possibility of communication with a particular audience, and a particular scenario of implementation.

The audience I have in mind is 'environmentalists' and the various organizations, societies, and movements concerned with environmental activism around the world. To keep my imagined scenario as concrete as possible, I focus on a particular group with which I have some experience and familiarity. In section one, I mentioned 350.org as an example of a growing environmental movement: here is a global community of people of good will who want to do something about climate change. Might there be some who, like Suzuki, sense the need for a larger cultural change, who sense that the old methods (i.e., protesting, petitions, public actions etc.) won't get us there? Are there some who might be open to looking at the broader problem? I am thinking toward the possibility of communication to the 350 community resulting in a small subgroup who are serious about tackling the larger cultural problem. How might those people be reached? What would be the content of that initial communication?

Some background on the group will help. There are over a hundred loosely affiliated local 350 groups around the world, supported (with online tools, resources, communications) by a small team of paid staff. Each group runs their own locally-driven campaigns, such as "fighting coal power plants in India, stopping the Keystone XL pipeline in the U.S., and divesting public institutions everywhere from fossil fuels."[19] Since 2007, 350 also has led national "Power Shift" convergence gatherings in countries around the world to develop new campaigns and strategies for action on climate change, and to equip participants to lead local action groups in their own communities.[20] All of this work aims to "dismantle the influence and

---

[15] "Time to Get Real on Climate Change," accessed July 3, 2014, http://vimeo.com/94622378.
[16] *Method in Theology*, 4.
[17] See William Zanardi's contribution to this collection at page 151.
[18] *Ibid.*, 153.
[19] "What We Do," 350.org.
[20] See http://globalpowershift.org.

infrastructure of the fossil fuel industry, and to develop people-centric solutions to the climate crisis."[21]

There is a clear reaching within the 350 movement for better, more *effective* solutions, and that is my concern also. So I imagine a project being pitched to the community: the creation of a small subgroup of collaborators concerned with finding new, better ways to "solve the climate crisis."[22] There is a diversity of possible means of communication, but for now I think simply of a single communicator, with some 350 staff members on board, speaking to the participants of a Power Shift convergence gathering—perhaps 1000 people. It might help if the communicator comes from within the 350 movement, or the "climate movement" more broadly, or is generally known and respected in those movements. It also will help to be clear that the proposed project is not meant to change the immediate form and direction of the movement—the small victories and successes are important right now, and current efforts to reduce carbon emissions should continue.

Instead, the proposal will be for a long-term project aimed at getting to the root of the climate issue, beyond the fossil fuel industry. The biggest challenge for the communicator will be to persuade the audience of the need to better understand the problem in order to come up with more effective solutions. That will mean looking beyond carbon sources and sinks to the economic, political, educational, social, and cultural aspects of decline that have led to the present ecological crisis. So, the communicator will be looking for volunteers, people from the gathering who are serious about taking on the challenge and willing to accept that results will be slow to emerge. Among the 1000, I imagine there might be 25 interested people who would sign up for the job.

Where would they start? I think our communicator could be quite helpful, not by directing the group, but by acting as a sort of mediator or facilitator. They might help bring the group to an initial consensus about the need for some of them to think about the historical question: what has been happening, up to the present, that has brought about this climate problem? Two or three people might look at economics, two or three might look at educational institutions etc. After a bit of lag time, others in the group could begin drawing on those results to move towards solutions. That way, slowly, there could emerge a preliminary division into past and future-oriented work. The collaboration would be messy at first, but the communicator could monitor things and give advice and suggestions about possible refinements in method, eventually bringing out a cycling of eight different tasks.

---

[21] "What We Do," 350.org.
[22] "350 Manifesto," 350.org, accessed July 7, 2014, http://350.org/350-manifesto.

IV. Concluding Reflections

The previous section is my beginner's attempt to think concretely about seeding functional collaboration in ecology. For now, the team of 25 collaborators is a nascent fantasy, and it calls for further fantasizing if it is to grow into something more. I have focused on 350.org, but I suspect that aspects of my imagined scenario might be applicable to other environmental groups and organizations. Might some of the ideas here also be applicable beyond ecology? Might SGEME, or similar organizations yet to emerge, make some use of these ideas in communicating and seeding functional collaboration in ecology and other areas?

I leave these questions for now to consider the feasibility and promise of the hands-on effort imagined in section 3. Does the project stand a chance? Its implementation would be a slow and challenging process. There will be the challenge of overcoming the desire for a "quick-fix" to climate change and other environmental problems. Because the dangers are immediate, there is a push for immediate solutions, and finding even 25 people in 1000 who would be willing to think about fundamental issues would be difficult. Furthermore, even with a functionally minded communicator providing some kind of mediation, bringing a group of 25 to a preliminary functional division of labor would be challenging simply because the move towards functional thinking in itself would be so novel.

So the project would certainly be "difficult and laborious,"[23] and not a solution for immediate problems. But without a new way, not only will we find ourselves fighting the same battles again and again, but we will be faced with increasingly dire situations as problems accumulate and decline accelerates. Our efforts to address immediate and isolated ecological crises, however successful in the short-term, cannot keep pace with the cumulative patterns of destruction. Such measures treat only the symptoms of an underlying cultural sickness.

Present-day economics is a major part of that cultural sickness that I have not yet discussed. It is, for example, well known that profit maximization is driving environmental destruction, and the project of section 3, were it to be taken seriously, would help bring out the need for a new science of economics. In his contribution to this volume, Bruce Anderson presents some of the most basic knowledge that would be part of that new science.[24] It also would become part of the worldview of a group like the team of "25." Though environmentalists and ecologists would not be expected to become economists, eventually some would need to know economics well enough to be able to collaborate with economists and contribute to the seeding and implementation of sane economics.

The larger hope is that the effort envisioned in section 3 may eventually seed, in part, a collaboration that is effective globally, in all fields, however patchy it may

---

23 *Method in Theology*, 4.
24 See Bruce Anderson's contribution to this collection at pages 121–123.

initially be. Perhaps we may even envision the kind of longer view expressed in McShane's "Arriving in Cosmopolis."[25] Might 100 groups of 25 gradually grow into a tower community of 250,000,000 by 9011 A.D.? Might we not imagine 250,000,000 Carers, "'living human bodies linked together … in the intelligently controlled performance of the tasks of world order"?[26] How strange that control will be is quite beyond our present fantasy: "a billion half-acre gardens, perhaps, with nano- and micro- and bio-mimetic technologies giving the average ten occupants of each garden a global intimacy and a local sufficiency?"[27] The particular effort envisioned here may or may not contribute to bringing about McShane's vision for 9011 A.D. But the present task of communication remains central and vital, if that vision of global care is to have a chance of being realized.

---

[25] Philip McShane, "Arriving in Cosmopolis" (keynote address at the First Latin American Lonergan Workshop, Puebla, Mexico, June, 2011), available at: http://www.philipmcshane.org/website-articles.

[26] *Insight*, *CWL* 3, 745.

[27] McShane, "Arriving in Cosmopolis," 9.

Philip McShane

This essay has six sections. The first section points to a creative jump regarding the third stage of meaning that ties in with the third line of Lonergan's spread of words on page 48 of *Method in Theology*. The second section focuses on new meanings of human evolution, mainly within general categories of meaning. The next section lifts our thinking into a focus on special categories. Section 4 turns to a single article of Thomas's *Summa Theologica* to illustrate the transition in theology involved in Lonergan's shift of the meaning of science. Section 5 talks of our stumblings in this volume. The final section muses about the road forward towards the distant goal — 9011 A.D.—of "Arriving in Cosmopolis."

1. A Third Stage and Third Line of Meaning

I might well consider this to be part two of an essay written by me in a summer month of 1961. Its title was "The Hypothesis of Intelligible Emanations in God."[1] It comes 53 years later than that first part: surely this is a record? Already there is a terrible density in the first part, a density requested by John Courtney Murray: put the message of the five *Verbum* articles in a compact form. But the density of that first part is just a modest effort of a first-year theological student who had yet to

---

[1] It was published in *Theological Studies* in 1962. I had moved, from teaching mathematical science in University College Dublin, into the study of theology, in the autumn of 1960, at Milltown Park, Dublin. That year I wrote "The Contemporary Thomism of Bernard Lonergan," published in *Philosophic Studies* (Ireland) in 1962. But the previous year I had offered it to Fr. Courtney Murray, who found it too philosophical but asked me to do an article that would present the core of the *Verbum* articles. Years later Fr. Crowe was amused by the story and remarked that Courtney Murray would never have done that had he known that I was in first year theology. I was to do the course on the Trinity in the academic years 1963-64, so I was invited to move to Heythrop College, Oxfordshire for that final year of theology. That eased the nerves of the professor in Milltown Park. I spent most of the fourth year of theology writing "Insight and the Strategy of Biology," published in the Lonergan *Festschrift*, *Spirit as Inquiry: Studies in Honor of Bernard Lonergan, S.J.*, edited by Fred Crowe, 2:3 *Continuum* (1964). Meantime I had published "Theology and Wisdom," *Sciences Ecclesiastiques*, 1963, "The Causality of the Sacraments" in *Theological Studies*, 1963, and "The Foundations of Mathematics" in *Modern Schoolman*, 1964. On the side I managed to get the usual S.T.L. degree in theology and arrange to avoid going to the Gregorian University for doctorate studies. This odd flow of events is relevant: as a scientist I was a displaced person in the theology of the time. But now, am I still a displaced person?

221

venture formally into Trinitarian theology. The density of the second is of initiating the venture into seeding a new Trinitarian theology after a half century of persistent climbing into the question, "What might being be?" Perhaps a gentler title for this second part might have been, "The Fact of Intelligible Emanations in God." The gentler title would hide from the reader, at least initially, the existential gap, the Beethoven pause, the Proustian weave.

But even the initial reading of that alternate title might halt you in the blunt odd word *fact*.

"What, then, is being?" is only a strategic twist on the road to the weave in molecular patterns of absolutely supernatural facts.

Thomas would have no difficulty with the word *fact*: indeed, he would have preferred the word *fact* in the title of the first part, my first essay. Would Lonergan have difficulty with *fact*, or indeed with any of the facts pointed to by either of my titles? I think not. He regularly delighted in my strange titles and now can only bodilessly grin at my suggestion that I am really only conjuring up a more factual version of the third line of his own odd spread of words on the top of page 48 of *Method in Theology*. The line reads:

"liberty   orientation, conversion   personal relations   terminal value"

Recall his comment in *Insight*: "In the first place, there is such a thing as progress, and its principle is liberty."[2] Liberty is there, emergent fact of the Rift Valley of Primitive Africa. In the last place there is terminal value, and no doubt you may think of it in terms of Aristotle's final causality, but here I am thinking of the fact of the friendly universe, and the claim of John in the name of Jesus, "As you, Father, are in me and I in you, may they also be one in us,"[3] the whole hominid hundred billion brained parade. So, in reality, liberty is Clasped, Embraced, and the fact of terminal value is a Hailing Cauling One.

But we must pause over this third line of *Method* 48, as I have done many times before. If one takes seriously the end of the second line then this third line doesn't belong. "What, then, is being?" It is the good of order.

But we are in this dicey domain of history, and the good of order is incomplete. Is it to be always so? A strange and worthwhile question about the everlasting; but here we think only of the tomorrows of the emergent good of order.

And thinking thus we can settle for line three being sneakily lurking in the previous two lines, a disturbing reality of pre-human and human history. And we can settle for Lonergan's version of it.

---

[2] *Insight*, *CWL* 3, 259.
[3] *John* 17:21.

But our conference and volume go further, and some of the collaborators are explicit about that further reach, that Cherishing in finitude which is the Hailing, the Caressing, that is a Divine speaker Calling.

The calling is the sequence of emergent goods, "always concrete."[4] And the sequence calls—is it not a part of the calling?—for its genetic conception in whatever fullness the human group can rise to "on the level of one's age,"[5] "on the level of one's time,"[6] but always everlastingly beyond, beyond especially the battered tadpole of our so-far wandering from the Rift Valley.

So, Lonergan is painfully led to his dream of the tower, a dream I claim to be lurking in that final line of *Method in Theology*'s word spread of page 48.

But might it also lurk in you?

That is the issue, the potential molecular and minding issue pointed to in you, in this volume. It expresses a Christian stand on the bold-faced question raised at the beginning of chapter 8 of *The Everlasting Joy of Being Human*: "**Do you view humanity as possibly maturing—in some serious way—or just messing along between good and evil, whatever you think they are?**" But the question is posed here in the fullness of tower-entry.

That fullness is captured in the words bracketed by *Embracing* and *Cauling*: *Luminously and Toweringly the Symphony of*. "The Symphony of"? We are back with the main message of the book, *The Road to Religious Reality*, and I do not wish to compact that meaning here other than to note now that the issue is the towering presence of the second Trinitarian Person between the third, Embracing, and the first, Cauling. What is at issue, the issue in you, is a cyclic taking off from the final words of Lonergan's systematic treatise on the Trinity: "joined to the Spirit in love and made living members of the body of Christ we might cry out, Abba, Father!"[7]

My footnote seven points to the identity of the text I used in 1961 and the present text, and it brings you to read freshly the title and its claim "Bernard Lonergan Evolved."

2. Evolving

How might I intimate that fresh reading? Some of you are members of SGEME, but perhaps scarcely recall what the letters of the word *SGEME* meant, quite apart from the serious meaning of the word that concludes the society's name: *The Society for the Globalization of Effective Methods of Evolving*. Evolving? It has at least a general meaning

---

[4] The first sentence of the chapter on "The Human Good," in *Method in Theology*, 27.

[5] *Method in Theology*, 351.

[6] *Ibid.*, 350.

[7] *The Triune God: Systematics*, *CWL* 12, 521. This is the translation give for the Latin of the previous page. That Latin is identical to the Latin of page 329 (Rome, 1957) of *Divinarum Personarum conceptionem analogicam evolvit Bernardus Lonergan* (translated: Bernard Lonergan evolved an analogical conception of the Divine Persons).

for you, but has it the beginning of the meaning, the mark, that Lonergan mentions in the Preface of *Insight*, the meaning of "the elapsed twenty-eight years" that "have left their mark upon me"? What might that mark have been on the fifty-year-old man?

More important—for the mark is a remote hidden molecularity—what might be your view of sharing that mark?

So I come to putter round the hidden molecularity of the word *luminously*. My co-author James Duffy refers in his essay to Teilhard de Chardin's book, *Science and Christ*.[8] Chardin, a marked man, writes powerfully as if he were communicating with somebody. He is not luminous about his failure to do so, in my sense of *luminously*. He is, of course, not alone in that. Non-luminous communication of serious personal depth is an axial thing, though you find oddities like Proust in various fields who tune into the sad reality of faded 18-year olds: "not old folk but young people of 18, very much faded."[9] Luminous communication is to eventually emerge, weaved into a HOW-talk that will, for example, eliminate the necessity of such appeals as Lonergan made at the conclusion of his *Verbum* articles about cultured reading.[10] When that language is an ethos of trail-presentation then its mark will be a shared global pattern, a marrowed sense of adult growth. Need I go on about that here? Indeed I do: only the rare person has the marrow marks, at any age, of the plodding climb to meaning, and the becoming stranger to themselves of last month.

But I do wonder now about any going-on here, puttering about the "commonly obscure gap."[11]

I am raising—am I really?—the question and the quest in you for a sensibility to the existential gap between today's horizon and tomorrow's, between any horizon and the field: "The field regards metaphysics as such, but the horizon regards metaphysics as possible-for-me, relevant-to-me."[12] Lonergan, after a quarter of a century searching, typed his fifteen-hundred-page climb on a little machine, and there was *Insight*, a printed handful in 1957. He was too busy climbing to talk scientifically about the climb, and this was true even of his last book, *Method in Theology*. He had, as he began that book in 1966, a powerful sense of the climb to which he was inviting his colleagues in theology, and he shared that sense and the concomitant gloom about the project with me that summer.[13] But we did not share—for neither of us had tried seriously to thematize it—a self-luminous comprehension of adult growth, one that was a formal comprehension. It is a

---

[8] Teilhard de Chardin, *Science and Christ* (London: Collins, 1965).
[9] Marcel Proust, *Remembrance of Things Past* (New York: Random House, 1932), vol. 2, 1042.
[10] See *Verbum: Word and Idea in Aquinas*, CWL 2, 223.
[11] *Insight*, CWL 3, 565.
[12] *Phenomenology and Logic: The Boston College Lectures on Mathematical Logic and Existentialism*, CWL 18, 119.
[13] In particular, what to put into chapter one baffled him.

problem of reaching and expressing the adult growth of a comprehension of everything.

> Now this comprehension of everything in a unified whole can be either formal or virtual. It is virtual when one is habitually able to answer readily and without difficulty, or at least 'without tears,' a whole series of questions right up to the last why? Formal comprehension, however, cannot take place without a construct of some sort.[14]

Lonergan's writing there is about getting a grip on the constitution of Christ. Our chat here is about something like getting a grip on our getting a grip on that topic, or any topic.

That formal comprehension is quite remote from present culture, indeed I suspect from the efforts of this next century. Heavens, we don't have a formal comprehension of the growth of a sunflower![15]

So what is to be effected by my short musing on the genetics of evolving a minding?

As I continue to weave round this topic—after writing this section and indeed section 4 below—it dawns on me that I am engaged in yet another essay within the formula of the volume. Even, accidentally, but with a cunning twist now, having the same numbering scheme.[16] Here, then, in section 2, is the discovery-content to be handed on. The discovery-content is my single line as a suggested replacement for the third line of Lonergan's spread on *Method in Theology* 48, and the luminosity of growing in its understanding.

Instead, then, of:

---

[14] *The Ontological and Psychological Constitution of Christ*, *CWL* 7, 151. The end of the paragraph is relevant to our musing: "Thus, if we want to have a comprehensive grasp of everything in a unified whole, we shall have to construct a diagram in which are symbolically represented all the various elements of the question along with all the connections between them."

[15] I had dabbled in the topic of adult growth in the 1950s, and later pondered over the works of Maslow and Aresteh, with a first serious blossoming of talk in the second chapter of *Process: Introducing Themselves to Young [Christian] Minders* (1989), but my better start of musing on the task was in 2001, in the end pages of *Lack in the Beingstalk: A Giants Causeway* (Halifax: Axial Press, 2006). A key nudge forward in the *Cantower* project of the next decade was the essay that became *Cantower* 2: "Sunflowers Speak to Us of Growing." On the problem of reaching a theory of botanical development, see "Method in Theology and Botany," Part One of *Method in Theology: Revisions and Implementations*, a website book.

[16] I leave this suggestion as I made it here, with the intention of turning my 6 sections into 4 by the device of sub-sectioning. But it would have been a pedantic move. Still, it is a nice exercise for the reader to see how the six sections might have fitted into the formula of four that dominates most of the essays in the volume.

"liberty   orientation, conversion   personal relations   terminal value,"

I offer:

Embracing Luminously and Toweringly the Symphony of Cauling.

But the subtlety of my offer is that the line is poised in a searching for a luminous towering meaning of personal relations, with little *p*s and capital *P*s. And in that offer there is in fact a set of lifts of the meaning of the search for personal relations, and here I only mention one facet of that set. There is a new twist to the prayerful reach for and attainment of Cosmopolis weaved into a reach for a genetics of asymmetric friendship.

The line points to the emergence of a metatheoretic consciousness that is to define humanity's climb in its personal intimacies and its phyletic glory, a defining that is to be an increasingly kataphatic luminosity, "so that joined to the Spirit in love and made living members of the body of Christ we might cry out, 'Abba, Father.'"[17]

3. Special Categories

The new turn expressed in those last paragraphs of section 2 loops out of my previous sketchings for this section. Yet musings on hand-on strategies for my foundational stand mesh with the large original project of talking about the challenge to face huge foundational discontinuities in theology. Fortunately James Duffy has already dealt with aspects of that challenge in taking up the topic of my five Cs that relate to the radiances in history of the Divine Persons: Clasping, Cherishing, Calling, Craving, Christing. He articulated aspects of the search for the meaning of *Cherishing*, that OM of the first Trinitarian Person that wormholes us all into a sort-of infinite finitude.[18]

The search for the meaning of that absolute supernature is, however, of the same ilk as the search of hypothetical normal humans for meaning in other zones.[19] But that natural normative search has been botched in various ways since the beginning of the axial period. The lift out of that botched state is one I identified, in my listing of 7 functional policies, as "The Tomega Principle."[20] I titled it Tomega

---

[17] Lonergan, *The Triune God: Systematics*, *CWL* 12, 521. These are the final words of Lonergan's treatise on the Trinity, the pointers to the beginnings of the post-axial praying that is to be both theology and common sense in the later stages of the third stage of meaning.

[18] The conclusion of my Epilogue homes in on this strangeness of finitude as a core venture in the theology of these next millennia.

[19] A context is "The Natural Desire to See God," *Collection*, *CWL* 4, 81–91. See note 21 below.

[20] The listing occurs in section 4 of *Cantower* 41, "Functional Policy," written in 2005. That *Cantower* ended the first surge towards 117 *Cantowers* (the number of *Cantos* written by Ezra Pound) and was followed by a long effort to understand *Method in Theology* page 250. Later

with a bow to Chardin's Omega point, and in doing so I recognize now that I was fiddling with the perspective of pure nature much as Lonergan did in *Insight*.[21] The Tomega principle is indeed borrowed from a slightly deceptive—certainly appearing empirically wobbly—perspective on human thinking that Lonergan proposes: "Theoretical understanding, then, seeks to solve problems, to erect syntheses, to embrace the universe in a single view."[22] Lonergan is flying high here, flying in the face of his sad comment on theologians: "Theologians, let alone parents, rarely think of the historical process."[23] But before I pick up on that and on what I call my fiddling, let me present you with my venture of establishing the Tomega Principle. I do so at length because it is important to see the *Cantower* struggle freshly, in this new context of the problematic of luminous growth.[24]

I focused a similar long effort at understanding content and context of that wonderful paragraph beginning, "study of the organism begins …" (*Insight*, *CWL* 3, 489), which ran to 41 essays, *Field Nocturnes*, another lead into the concluding *Cantowers* which finally reached the number 158 (117 + 41).

[21] See note 19 above. Lurking over my Epilogue are questions about the nature of Christian philosophy, questions I have entertained over decades. I think of Lonergan's review of books on the topic in *Shorter Papers*, *CWL* 20, 222–23, which ends marvelously thus: "I am led to suggest that the issue which goes by the name of a Christian philosophy is basically a question on the deepest level of methodology, the one that investigates the operative intellectual ideals not only of scientists and philosophers but also, since Catholic truth is involved, theologians. It is, I fear, in Vico's phrase, a *scienza nuova*." A simple answer can talk of a *de facto* Christian philosophy. *De facto*, functional collaboration emerged out of a problem in Christian theology, even though, as various contributors here note, it is a problem that ferments forward out of other areas. Again, *de facto*, the maturity of analyses of minding, with its need to discover the meaning of "is? is! is.", is a Christian achievement, even if verification in science nudges towards a methodological answer. One should note here that a lot of inter-author dialogue of Lonergan studies does not advert sufficiently to the fact that Lonergan's study of mind in *Verbum* is way out of the ballpark of contemporary philochat, clustered round various road-houses, where "the halfway house is idealism." *Insight*, *CWL* 3, 22. And finally, there is the issue of personal relations at the heart of this essay. Note 1 of Lonergan's Epilogue to *Insight* is quite insistent about the strange real-world context of such hello-saying.

[22] *Insight*, *CWL* 3, 442.

[23] "Finality, Love, Marriage," *CWL* 4, 47.

[24] I do hope you catch the humour of this statement. What "length" is required "to see"? I look back myself at the length of more than a decade reaching to now, now see freshly. How do I get you to see? I think now now now of the leads in one of my favorite books, Rita Carter, *Mapping the Mind* (London: Phoenix, 2002) about "the nuts and bolts of thinking." *Ibid.*, 312. Really seeing, getting these insights seeded and sown? "The process of laying them down permanently takes up to two years. Until then they are still fragile." *Ibid.*, 268. Furthermore, the permanent theoretic laying down that is our topic is quite

I come now to [4], the Tomega Principle. You will find the principle formulated for the first time in my writings in *Cantower* 4 of July 2002, although it was part of my dynamic from ... way back! And perhaps my drawing attention to this, by inviting you to pause over this, will help you to "handle," hearthold, get your molecular head around about about about this doctrine and these doctrines.

The Tomega Principle is printed out on page 7 of *Cantower* IV, and I read now my comments there. I meant just to quote the definition and move on, but, my my, that page was worth my reading again for the first time with its burst of fresh meaning nudging me along my dark galactic trail. So, I will type the whole page in here, thus talking to myself again, beginning again to taste the tease of Lonergan's marvelous leaf 417 [442] of *Insight*. To think that I missed the key pointing of it in my readings of forty years! So: let me give you the beginning of section 1.2 of that *Cantower*, titled "A Pert Direction."

What we are reaching for, THEN, is a can-tower self-luminosity of molecular intelligence implementing its explanatory self-tasting in an efficient spin-in and spin-off of noo-feedback.

There you have it, in foundational fantasy, but not yet in doctrinal bluntness.

Here, then, you have a pert, saucy, attempt at doctrinal bluntness. That gives you one of my senses of *pert*. The dictionary may also give you PERT, initials for Program Evaluation and Review Technique, and that also pertains here. But the central meaning is the naming of Candace Pert.[25]

I am not settling here into a particular functional specialization: indeed the *Cantowers* in general can be read as popularizations, literary invitations, C$_{59}$,[26] pointing towards the later hodic adventures. But it may be as well to be saucy up-front with a metadoctrinal statement of Lonergan that I make my own. Let us isolate it boldly, titling it *Tomega*.[27]

---

different from the laying down which results from habits of scholarly consciousness which can leave you in "no man's land." *Philosophical and Theological Papers 1958–1964*, *CWL* 6, 121.
[25] Candace Pert, *Molecules of Emotion* (New York: Touchstone, 1999).
[26] See *A Brief History of Tongue* (Halifax: Axial Press, 1998), 108, for the relevant matrix. The "9" signifies that the communication reaches beyond the matrix of collaboration: see the diagrams there on pages 109 and 124.
[27] 'To Omega' brings to mind, perhaps, Chardin's vision of an Omega point. But I

Tomega: "Theoretical understanding, then, seeks to solve problems, to erect syntheses, to embrace the universe in a single view."[28]

This sentence begins a powerful paragraph, a powerful stand, against commonsense eclecticism. Only a few years ago I began to grasp its significance as a foundational statement, a statement of general categorial orientation relevant to all human inquiry and life, a claim that goes counter to an accepted culture of specialization, a consequence of the fact that organisms live in a habitat but the human organism lives in the universe. Furthermore, in these last few years, the sentence has been further lifted, embraced, molecularly braced in a self-mediation—like a luminous watch[29]—by work that merges with and transposes the efforts of Candace Pert. And now I read, with fresh strange ayes, the last paragraph of my effort of 1989: "The third stage of global meaning, with its mutual self-mediation of an academic presence, is a distant probability, needing pain-filled solitary reaching towards a hearing of hearing,[30] a touching of touching, 'in the far ear,'[31] 'sanscreed,'[32] making luminously present, in focal darkness, our bloodwashed bloodstream. It is a new audacity, a new hapticity, to which we must aspire, for which we must pray.'[33]"[34]

I have thus quoted at some length to let you sense the years of climbing beyond that final quotation from *Process*, a book that focused for a year on the problem of handing on. It was, of course, like many of my works, rejected by publishers, and the hand-on message and appeal remain, till now, a dangling baton in the human race.

---

have in mind also Aristotle's view of the finest life, and Thomas' view of human happiness, and Lonergan's view of the significance of leisure, and my own view of the radical failure of contemplative traditions East, West, and South.

[28] *Insight*, *CWL* 3, 442.

[29] The implicit reference here is to Lonergan's discussion of the mediation of Christ in prayer, where he moves up through analogy with the workings of a watch.

[30] "Merced Mulde" "Yssel that the limmat?" *Finnegans Wake* (New York: Viking, 1939), 212, line 26; 198, line 13. The strange reduplicative process is the central drive and fantasy of this *Cantower*.

[31] See John Bishop, *Joyce's Book of the Dark: Finnegans Wake* (Madison, WI: University of Wisconsin Press, 1986), 343–46.

[32] *Finnegans Wake*, 215, line 26.

[33] Philip McShane, *Process: Introducing Themselves to Young (Christian) Minders*, 1989, 162–63. The notes internal to this passage are from the original.

[34] I am quoting here from *Cantower* 41, "Foundational Policy," 14–16, with an inner quotation from *Cantower* 4, "Molecules of Description and Explanation," 7–8.

But I include it here with starling fresh glimpses of what was wrong in my first appeal in Florida for a "third way, difficult and laborious."[35] I was poised then as a scientist,[36] but the dominant ethos of the group was the ethos of a humanities molassed, mole-assed, in "academic disciplines."[37] I brooded over the failure of that conference, its papers and discussions, as I took flight back to Ireland with the task of editing the 72 disciples. Now I see a little better the trail of non-science after Thomas, the wonderland of name-droppings in philosophy and theology and in general literate and literary criticism. I think now of Stephen J. Joyce, James Joyce's grandson, remarking (the odd language is his), at a 1986 conference of Joyceans in Copenhagen, "if my grandfather was here, he would have died laughing." But the devouring of Lonergan's precise metascience by academic name-droppings is no laughing matter: it has become a vulgar immorality.

What, then, to do about handing on? I have drawn a parallel between the non-science that followed Thomas and the present state of Lonergan studies.[38] Why not try a little venture into a corner of Thomas that connects that corner with present needs? So: we have the following section. And might not a further digestive or indigestive intake of our stumblings here help? So, there is here section 5. And there is always the other type of venture to which we turn in the final section, the venture noted by Lonergan at the end of his essay on "Healing and Creating in History."

Is my proposal utopian? It asks merely for creativity, for an interdisciplinary theory that at first will be denounced as absurd, then will be admitted to be true but obvious and insignificant, and perhaps finally be regarded as so important that its adversaries will claim that they themselves discovered it.

4. A Simple Illustration

My simple illustrating hovers round the question raised by Thomas: "Whether the gifts of the Holy Spirit are conveniently counted as seven?"[39] but it might be more interestingly put as the question, "How does verse 2 of *Isaiah* chapter 11

---

[35] I am repeating, of course, words ending the third paragraph of *Method in Theology*. At the 1970 "First International Conference on Lonergan" in Florida, the book was not yet around, apart from its fifth chapter, "Functional Specialties," which had appeared in *Gregorianum* 50 (1969). I had the advantage of being tuned into this "third way" by Lonergan in the summer of 1966, and I had fleshed out the need in musicology for such functional collaboration in one of my two Conference papers, "Meta-music and Self-Meaning."
[36] See the conclusion reached at the end of note 1 above.
[37] *Method in Theology*, 3, the last two words of the first page of the book.
[38] In the Prologue to *The Everlasting Joy of Being Human* (Vancouver: Axial Publishing, 2013).
[39] *Prima Secundae* q. 68, a. 4.

conveniently hit the streets of New or old York, of Berlin or Beijing?" If you stay in the mood of our musings about evolving in section 2, you might, indeed should, have your focus on the question, "How does this text weave around my bones?"[40]

The issue of the text? The reading by you now of *Prima Secundae* q. 68, a. 4. What I would have you begin effectively to grasp is that you may not be, indeed, as a human, are wildly not,[41] up to any serious scratch with the first rule of reading, "understanding the object."[42]

But let us start with the problem of having the text, so off we go to *Wikipedia: seven gifts of the Holy Ghost*, to find Thomas' text, the Vulgate, lined up with Hebrew, Greek, and English versions. We already have problems crossing the page: how did *pietas* get in there? We need research help, indeed functional research help.

But back to the first rule from *Method* we go, for deeper discomfort. What is the object, and when is it? *When* seems a crazy twist, does it not? Yet it was in Thomas' mind, chasing forward from the previous question 67, all of which deals with faith, hope and charity after death,[43] and in question 68, article 6, he has a blunt response about the gifts "in patria": "they are to be most perfect"—"they"? They? Are we not caught here in that new third line of *Method in Theology*, 48, with Gift and Givers dancing with us in molecular patterns that are a clutch of terminal value? And might you not now resonate with that previous Epilogue of mine, "Being and Loneliness," which begins with Hermann Hesse's "... each member, each group, indeed our whole host and its great pilgrimage, was only a wave in the eternal stream of human beings, of the eternal strivings of the human spirit towards the East, towards Home ..."[44] and ends with the words "Infinite Surprise"?[45] And please, please don't mistake my slim understanding **there** of forty years ago with my shabby

---

[40] I hardly need at this stage to note the normative operative presence of the strategy of generalized empirical method as described at the top of page 141 of *A Third Collection*, edited by Fred Crowe, S.J. (Mahwah, NJ: Paulist Press, 1985).

[41] We are dealing with our absolutely supernatural reality, everlastingly and joyously elusive in *patria*.

[42] *Method in Theology*, 156. It is the section title.

[43] A discomforting pause is called for here, perhaps beginning with the expressed light-weight musings of Lonergan on faith, hope, and charity in chapter 20 of *Insight*. Then there is the discomfort of noting a prerequisite for seriousness in the "sixty three articles in a row" (*Grace and Freedom: Operative Grace in the Thought of St. Thomas Aquinas*, CWL 1, 94) that are the context of thinking out virtues and gifts. But beyond that I would note the lift the whole enterprise gets from bowing to the molecularity of virtuous activity. It is a neurodynamic achievement of flexible circles of ranges of recurrence-schemes. For a start here see *Quodlibet* 3, "Being Breathless and Late in Talking about Virtue."

[44] Hermann Hesse, *The Journey to the East* (London: Peter Owen, 1970), 12. Quoted in my *Wealth of Self and Wealth of Nations: Self-Axis of the Great Ascent*, 101, at the beginning of the Epilogue.

[45] *Wealth of Self and Wealth of Nations*, 111.

better grip now[46] on this present writing subject's marrow-minding harrow-blading "the stooks rise around"[47] the subject-as-subject reaching towards all and all of us and Those Subjects-as-Subjects.

But why not lift that "don't mistake" to a "HOW[48] not to mistake" achieved in some serious communal fashion? Knowing the object turns out to be "our whole host in its great pilgrimage" stumbling towards a heuristics of the object and a never-luminous sense of "all that is lacking."[49] The stumbling takes shape as a sliver of "a new front-thesis on the mystical body, that front-thesis eventually to be integrated in the sublated genetic systematics of all such theses through the ages."[50] The pre-sublated genetic systematics is the content of the meaning of **Comparison**, the center-piece of the strategy of the fourth functional specialty. The sublation consists of a community, in four situation-rooms, picking up on the "level of the times"[51] nudges towards refinements. In Thomas' time such nudges would be a better version of his 6 pointers against the convenience of sharing his way of thinking about the gifts of the Spirit.[52] In our time such pointers might come out of amygdalic considerations of the patterns of response, in the full heuristic of their chemo-dynamics: that heuristics also being a front-system of a genetic grip on the genetic dynamics of the integral human operator.

I slipped past, in that dense paragraph, the vast difficulty of present Lonerganesque imagination in thinking of the present cyclic operation of any pre-sublated genetic systematics. Such pre-sublation, at a first level, refers simply to the Standard Model in any discipline. I borrowed the notion of Standard Model, perhaps a decade ago, from contemporary physics. It is what is assumed to be operative in the Tower labours of any science, and, under internal strain, in revolutionary efforts in that science. But it is important to recognize the operations at what I might call

---

[46] I think of my three books of 2012–13 mentioned in these notes, but would draw attention especially to the climb represented by the twenty-one *Posthumous* essays. But one must note as still unexpressed the refined sublation of the canons of hermeneutics that are hinted at here throughout this section. Indeed, the core pointing is towards the need to conceive the luminous towering heuristic of what I call 60910: the paragraph that crosses from *Insight* 609 to *Insight* 610.

[47] From the first line of Gerard Manley Hopkins's "Hurrahing in Harvest," a poem that gives "lift up heart, eyes" (line 5) to our enter prize.

[48] It seems important to draw attention to that larger project of a luminous methodological language, a language which strains to establish a Home Of Wonder.

[49] *Insight, CWL* 3, 559.

[50] Philip McShane, *The Road to Religious Reality* (Vancouver: Axial Publishing, 2012), 38.

[51] *Method in Theology*, 350.

[52] The six pointers are the usual *videtur quod non*—"it would seem not"—that was part of Thomas' strategy. How such pointers weave into the first four specialties is not a simple matter.

a secondary level, indeed suitably so named, since I am talking, though loosely, about the difference between primary relativities and secondary determinations.[53]

This puzzling can carry us, in the present section's musings about reading, into the task of reading **to**. The reading **to** is circumstantial, and, taken in its fullness of reading **to** and **into** it, must to thought of as effective at least within some normal law statistic. That **reading to and into** is intrinsic to the science, any science, in the new ethos of which I write.

One may continue my paralleling with physics and think of Maxwell's equations, primary relativities that may be gradually recontextualized or even replaced, but advances can occur—and often are a core of doctorate work—where possibilities of significant secondary determinations emerge.[54]

As I write it becomes only too clear to me that I am rambling round a large book,[55] indeed perhaps a large book simply on the eighth functional specialty in the dynamics of its recycling through the first specialty of successes or failures.[56] But the rambling does fit into the purpose of this short section, which points to the complexity of the replacing of Thomas' effort. We have touched on problems of functional research, interpretation, history, and the consequent strategy described on *Method* 250 that would replace, in this case, Thomas' six pre-puzzlings. But these puzzlings are followed by the "*Respondeo dicendum*," which calls, in our new context, for the large complexity of talking forward, bringing in further twists on the meaning of *reading to*. One may think of Thomas' response as fitting in with Lonergan's optimistic sketching in *Insight*: "addressed to an audience that similarly grasps the universal viewpoint," though the dynamics of UV + GS was way beyond Thomas' imagination. The work of addressing, as Lonergan found ten and a half years later, needs to be split into a sequence of audiences, a bucket-brigade, if it is to be effective.

---

[53] See *Insight* chapter 16, section 2 and *The Triune God: Systematics, CWL* 12, Appendix 3.

[54] This is a well-known strategy in doctorate theses in the full range of sciences and indeed humanities: one brings to bear an established theoretic—primary relativities—on secondary or boundary conditions. See the following footnote.

[55] In the final section I write of the road forward, and mention thesis-writing. Here I note a neat thesis-topic: "Functional Specialization and Question X of Thomas *Summa*." Any question would do, indeed, as here, a single article is enough. I have skimmed over suggestions about that large book in the present case. There are a host of foundational questions as well as the functional structuring that I hint at in the text. I cannot, however, resist noting the sublation of the meaning of "fear of the Lord" that occurs in a full subject-as-subject sacrament-of-the-present-moment ingesting of (discernment)[3].

[56] This is a hugely important aspect of the re-cycling process. An achievement of a village or a campus cycles round and in the process gains potential secondary determinations that may fit all villages, all campuses. But the fit can fail, and that failure is a source of enlightenment for the later rounds of cycling. I would note here, in passing, that the "which have other grounds" of line 12 of *Method* 250 are not spun off but swing in especially to the situation rooms of the eighth specialty.

There are eight situation rooms towering over the global situations of classrooms, banks, churches, parliaments, where the gifts of the Holy Spirit need airing and chemicalizing,[57] and "such situations are the cumulative product of previous actions and, when previous actions have been guided by the light and darkness of dialectic, the resulting situation is not some intelligible whole but rather a set of misshapen, poorly proportioned and incoherent fragments."[58]

Have we not a new context for musing over the claims of the final page of *Method*'s chapter on "Meaning," about "educated effete,"[59] that concludes with the much-corrected final sentence, "Never has the need to speak effectively to undifferentiated consciousness been greater."[60]

But mainly my effort here has been to give an added hint about the nature of the great leap from Thomas to a "third way, difficult and laborious"[61] that leaps beyond the alchemy of "academic disciplines."[62]

5. Our Stumbling Efforts

Have we not now freshened the context for musing over the stumblings of this volume? The "fresh third line" suggested, presented, by the Epilogue's title is a piece of the foundational lift I have been aiming at in these past five years. Musing over that lift now in relation to the seeding problem should help all of us to envisage in a practical beginners' fashion the tasks that face us in this leap beyond alchemy.

The key piece that I have selected for our helpful musings here is the short piece by Lonergan titled "The Genetic Circle."[63] I quote the entire piece.

> That circle—the systematic exigence, the critical exigence, and the methodical exigence—is also a genetic process. One lives first of all in the world of community and then learns a bit of science and then reflects, is driven towards interiority to understand precisely what one is doing in science and how it stands to one's operations in the world of community.

---

[57] There is, for instance, a serious literature regarding amygdalic responses inviting us to intussuscept "fear of the Lord" in a homely manner that can poise each of us to admit, into the sacrament of the present moment, a chemo-verbal self-address, "of course, I continue to be a messer." The messing, of course, is in the reality of our drift since the Rift Valley.

[58] *Method in Theology*, 358. Lonergan adds a footnote here: "On this topic see *Insight* pp. 191–206, 218–232, 619–633, 687–730." The page references are to the first edition of *Insight*, trackable in *CWL* 3 without difficulty.

[59] *Method in Theology*, 99.

[60] *Ibid.*

[61] *Ibid.*, 4.

[62] *Ibid.*, 3.

[63] *Early Works on Theological Method I*, *CWL* 22, 140. I note that the piece belongs to 1962, so prior to his discovery of the functional structure of collaboration. The piece has already been commented on by me and others in this volume.

And that genetic process does not occur once. It occurs over and over again. One gets a certain grasp of science and is led onto certain points in the world of interiority. One finds that one has not got hold of everything, gets hold of something more, and so on. It is a process of spiraling upwards to an ever fuller view. That circle—systematic, critical, and methodical exigence—does not occur just once. It occurs over and over again in the self-correcting process of learning.

The editors' Introduction to this volume has already weaved successfully round the various stumbling contributions and their significance. I move on here by noting how appropriate some of Lonergan's words here are to our efforts in this volume. "One gets a certain grasp of science and is led onto certain points in the world of interiority. One finds that one has not got hold of everything, gets hold of something more, and so on." This Epilogue was made available to the contributors prior to our Vancouver Conference, prior to the opportunity for revisions in the light of the conference, and I asked the group not to take into account the lift of foundations pointed to in my Epilogue: it was difficult enough to gather ourselves, as it were in random dialectic, round the contributions of the papers and that week's discussions. My Epilogue is more about the full road forward, and the place of the title of this Epilogue in that road in Christian theology is to be my final topic in the next section.

So let us reflect with energetic and creative imagination on the "certain grasp of science," the "certain points in the world of interiority," the shaky hold on everything.

Most obviously, our efforts were scattered, dippings into various disciplines, expressions of "certain points" that were regularly not original, not fresh lifts to the cycling of our non-existent science, not related in any obvious way to one another. We were doing exercises in order to find our way into "The Genetic Circle" that Lonergan did not envisage for three more years after the quoted piece of writing. But note that he was talking unequivocally about a science, although he was well aware of stumblings and beginnings: "there is a genetic process from the world of community to the world of theory."[64] So, we and you notice, in our efforts, that we stumbled away, as best we could, from the ethos of academic disciplines. We pretended to be "at the level of the times,"[65] as any wise doctorate student does in a doctorate thesis. But none of us were. Further, part of the paradox of luminosity and adult growth is that elder members of our group were regularly better tuned to "all that is lacking" than younger members. I, then, more than others, knew what a shabby shot we were having at getting the show on the road. I had, especially, learned a great deal about the needed shabby start from decades of mucking

---

[64] *Ibid.* In the next section titled *Aberrations and Deviations.*
[65] *Method in Theology,* 350.

around.[66] I point to one instance that should encourage: my struggle with Fred Crowe's gallant efforts. It took me some years to get from my struggle with his work to a beginner's grip on the character of the functional researcher.[67] We—and many of the writers in this volume were involved—struggled towards a more refined grip on that character in us through the first *FuSe* seminar, an enlightening process that revealed to us just how far we and Lonergan studies generally were, for example, from the parts of the standard model sketched by Lonergan in *Insight*.

Now the marvelous thing about this failure to rise to *Insight*'s challenge is that "The Genetic Circle" shifts it from center-stage. The shabby start we need, "the certain grasp of science," is a grasp of the cyclic collaborative character of the work. So, we read Lonergan's paragraph differently fifty years after he wrote it. We start in the world of community, a community that has not clicked to his leap to a new science. So, this little group of writers "learns a bit of science and then reflects, is driven towards interiority to understand precisely what one is doing in science and how it stands to one's operations in the world of community."

The reflection begins now, after our seeding effort, and it is a reflection that we wish to share. What is functional collaboration? This volume and its conference is just not it. What is it to ask, "What is functional specialization?"? It is to ask a scientific question that has little data:[68] so we are thrown forward into creative heuristic boggly-genetic imagination backed by our stumblings. "One gets a certain grasp of science and is led on to certain points in the world of interiority." In the conclusion of the next section I will turn to the point of interiority represented by my title. But here I wish, plead, pray, for our focusing on the need within globalization on an interior visioning that is bent, Toweringly, on eliminating totalitarian ambitions. "That's a fundamental concern of method, eliminating

---

[66] Still with me, vividly, is the summer moment of 1969 in the Bodleian library when, looking at the shelves of the collection on musicology, there issued the gospel, "this is the way to the musey room." *Finnegans Wake*, 8. The result was "Metamusic and Self-Meaning," a paper for the International Florida Lonergan Conference of 1970 on functional collaboration as needed in musicology.

[67] I refer mainly to Frederick Crowe, *Theology of the Christian Word: A Study in History* (New York: Paulist Press, 1979). I worked on it pretty-well since its appearance, but I tackled its character as functional specialist history a decade ago, in section 4 of *Cantower* 38, "Functional History," with fairly negative results, apart from noting its brilliance as a pedagogical work. Later I tackled the book again, in the five *Humus* Essays, 8 – 12, and began to see its role as functional research, which led me to views expressed in the first ten *FuSe* Essays, numbered 0 to 9.

[68] My *Method in Theology: Revisions and Implementations* (2008) deals with this issue in chapter 10, "Metaphysical Equivalence and Functional Specialization."

totalitarian ambitions."[69] The concern is as old as Paul's *First Letter to the Corinthians*, and my two points—here and at the end of the next section—echo neatly the need to bracket I *Corinthians* 13 between I *Corinthians* 12 and I *Corinthians* 14. I have said enough, over the past 45 years, about that bracketing. I think now of the wish and prayer of my native Gaelic, 'Go neirigh an bothar linn,' may the road rise with us. And of course it will, within an emergent probability of the absolutely supernatural. But that absolute nudges us to change what is bracketed, I *Corinthians* 13, the Clasp of the Big Bang, into a scientific symphony in our bones: eventually, acceleratingly, within an effective slick click clock cluck Clutching recurrence-scheme of Remembering of the Future.

6. The Road Ahead

It might well be that—ho, ho me being dead—would help the road rise, to recall again my Gaelic phrase. But let us not bank on that, surely a rather trivial motivation to "slip away before they're up."[70] I do hope, however, that my musing here will ground cynicism about whatever laundering laudations occur at various conference centers on my demise. Unless, of course—"Ho hang! Hang ho! And the clash of our cries as we spring to be free"[71]—they include effective repentance and apologies not just to Lonergan but to "the order of the universe" and "that order's dynamic joy and zeal."[72]

I look now on the present narrow fixity of Lonergan studies with growing horror at its damaging of the deep and long-lasting progress envisaged by Lonergan. The damaging is most evident in the misdirection of these next generations of students, who are being steered into "academic disciplines" rather than into the "third way, difficult and laborious" that is the heart of Lonergan's revolutionary thinking.[73]

I have, in these past few years, very deliberately moved to a limited out-spokenness, even adding the satire of the name "The MuzzleHim Brotherhood." It is a limited out-spokenness, slipping past the task of pointing in constructive criticism to the trapped scholarliness of most of the present Lonergan leadership. But it is of considerable value for us to pause now over the character of that task.

Let me be bluntly clear. The task is not one that fits into the cycle of functional collaboration as skimpily illustrated by section 4 above. It is a task that is to be the

---

[69] "An Interview with Bernard Lonergan," edited by Philip McShane, in *A Second Collection*, edited by William Ryan and Bernard Tyrrell (London: Darton Longman and Todd, 1974), 213.
[70] *Finnegans Wake*, 627.
[71] *Ibid.*
[72] *Insight*, CWL 3, 722.
[73] I surely need not reference again those beginning paragraphs of the first chapter of *Method in Theology*, that scream of our gracious freedom.

fruit of an undeveloped eighth specialty, or if you like what I talk of as C9. To think of it otherwise, and to act on that thinking, would prolong the present muddlings.

A parallel should help. There are the present muddlings that belong to the pseudo-science of economics. How are we to break forward from them? [a] by pushing for elementary reforms of the beginnings of economic education in the later grades of school and in popular media; [b] by serious efforts to identify and eliminate the destructive contribution to recent history that is the mix of trading with the foibles of commodity money transactions, a galloping illness of at least the past eighty years.

I recommend that those interested in what Lonergan suggested in his life-commitment give time both to the parallel in economics to what it parallels in Lonergan studies.

Note, first, that [a] and [b] in economics involve communications in the sense I write of as C9. In its maturity it is to be the fruit of the eight specialties, but in its initial stages it is be a matter of generating a popular ethos regarding some elementary disorientations, not at all an easy task. Helpful here, of course, are the reflections of Bruce Anderson and Michael Shute in the present volume, as well as their and my own past efforts.

But best stick here with the paralleling of this with the task of shifting theology, and in particular Lonergan studies, into its proper role, "a *regina scientiarum*, not merely a constitutional monarch,"[74] or someone fiddling in Rome. Perhaps there is a problem here of a new name rather than the old wine bottles of *theology* and *philosophy*: so I have written of *futurology*.[75] I would not have us distracted by my suggested name. The important thing is the change of popular ethos in theology that is to parallel and twine with the shift in economics.[76] Something of the strategy is captured in the first part of a book footnoted immediately, *An Introduction to Modern Economics*—a fairly elementary ramble through the story of economics that ends with the claim, "It is time to go back to the beginning and start again."[77] And perhaps there is something parallel to this ramble and conclusion in the more complex

---

[74] Lonergan, *Phenomenology and Logic*, *CWL* 18, 126.
[75] The search for a new name of global concern has occupied me for a decade. *Futurology* seems to fit the bill. I would note that it especially challenges philosophers and theologians that emphasize merely viewing the past in scholarly fashion, or such a fashion as to get no further than taking some position on what is real, without a serious facing of the future. In Lonergan studies, for instance, one finds little serious work on the forward specialties.
[76] The twining, apart from functional twining, is a deep twining around a fuller concrete meaning of *promise*, which would bring together the meaning of money and the promise of the New Testament.
[77] Joan Robinson and John Eatwell, *An Introduction to Modern Economics* (London and New York: McGraw Hill, 1973), 52.

ramble of my early seminars,[78] where there emerged in the first seminar—on Functional Research—an identification of the Lonergan school's neglect of "fruitful ideas."[79]

Scattered pages of my scribbles about strategies that would help to remedy that neglect nudge me now, not to write on, but to leave those leaves drift into dust.

So: "my leaves have drifted from me. All. But one clings still."[80] "Just a whisk brisk sly spry spink spank sprint of a thing theresomere saultering."[81]

My title, then, THEN[82]: *Embracing Luminously and Toweringly the Symphony of Cauling*. Surely theresomere is a saultering of the future? Forget the Tower stuff if you must, then, THEN, and think only of the Bower. "I will build my Love a Bower"[83] is the song of Sufi and Zulu. But there can be that inner singing of globalogians that weaves the singing into science and lays down loggias for piccolo peeks and peaks. The singing can and will Gracefully and globally marrow-mindedly Sonflower into the Father's Cauling so as to turn time slowly into the Middleman's

---

[78] This was an ambitious project of on-line seminars cycling round functional collaboration. They were to be 25 seminars, each lasting 3 months, documented in a proposed cluster of well over one hundred *FuSe* essays. There are about 35 *FuSe* available on-line at: http://www.philipmcshane.org/fuse.

[79] *Insight*, *CWL* 3, 254; 264.

[80] *Finnegans Wake*, the final page, 628.

[81] *Ibid.*, 627.

[82] I end with brutal brevity, yet the brevity is alleviated by the reference to—and of course, an invitation to read—*Cantower* 5, "Metaphysics THEN." This was the *Cantower* in which I bade farewell to Pound's *Cantos*, winding my musings into the old Scottish song about building a bower (see the next note). I paused over the notes to his final 117th *Canto* and asked, "what are we to make of the closing rhythms as he climbs to the last of his 800 pages while we envisage a love-bower shared, encircling and encircled, circumincessed?" *Cantower* 5, 24. Haunting the *Cantower* and, indeed, this whole epilogue-enterprise, are those strange final lines, his last poem, from Samuel Beckett, quoted at the end of the first footnote in *Cantower* 5: "go end there / where never till then / till as much as to say / no matter where / no matter when."

[83] This is the title of the first section of *Cantower* 5. The other sections are titled "By Yon Clear Chrystal Fountain"; "And All Around the Bower"; "I'll Pile Flowers from the Mountain." **All around the bower** was to be the meaning of the diagram within section 3, named later *W₃*, sensed still later as finite cravings' **Double You Three**. The diagram points to the full task, a task pointed to in the previous note by the quotation from page 24 of *Cantower* 5. The contemplative lover needs, e.g., to climb through *Insight* 15 and 16, or Appendix 3 of *CWL* 12, to begin the climb to adequate intimacy with the Three. James Duffy's paper points to a beginning, and Michael George's venture into "craving" ("Finality, Love, Marriage," *Collection*, *CWL* 3, 49) opens up the issue of the presence of sexuality in the Bower.

tuneblood, not a mythic turn but a cyclic turn of finitude's Trinitarian meaning Within and In and Towards truth, for "the real issue, then, then, then, is truth."[84]

---

[84] *Insight*, *CWL* 3, 572, slightly wormholed.

APPENDIX: RESCUING SEXUALITY

1. Context

In a concluding episode of the British television series *House of Cards* Ilsa Blaire reveals wonderful fifty-year-old breasts, both prone-poised and swaying in apparent sexual frenzy, and her acting partner, Paul Freeman, displays his nicely-aged arse. Have we not here a Christian problem, in the performance, in the watching? Is it not, to say the least, naughty to be watching such a naked display of sexual delight? Moreover, should we not fault the performers, even if the frolics are fiction? The fictional characters are caught in adultery, and the performers' lives are surely pushing beyond the edge of decency. But let us lift this further into what we can certainly consider higher Christian realms, the realms of St. Ignatius' *Spiritual Exercises*, where he presents us in the conclusion with the challenge of "Contemplation for Obtaining Love." Is it quite beyond our fancy to include the sights and tastes and smells portrayed by those actors in our reach to thus obtain a glimpse of the divine?

The issue raised by the last remark might be associated with the parallel problem that Matthew Fox discusses in the 25th chapter of his book, *The Coming of the Cosmic Christ*.[85] The chapter is titled "The Cosmic Christ and a Renaissance of Sexual Mysticism." Read it and weep. But here I am not writing about mysticism but about ordinary interpersonal living with the wondrous allure of the Divine.[86] We can be led back, by Fox's musings, to sniff out the tradition lurking in the penumbra of the word *naughty* of the first paragraph, a tradition about which Bernard Lonergan puzzled in the concluding pages of his essay of 1943 on "Finality, Love, Marriage."[87]

---

[85] Matthew For, *The Coming of the Cosmic Christ. The Healing of Mother Earth and the Birth of a Global Renaissance* (New York: Harper and Row, 1988). I remind you that this little essay is an appendix to my Epilogue to *Seeding Global Collaboration*, which I aimed to keep within a few pages. So, I develop little here, but perhaps I help to stir the healing and the renaissance of which Fox writes.

[86] "Allurexperiences," all your experiences, is the topic of the central *Posthumous* Essay, central in both senses, being the 11th of twenty-one essays moving round the present topic.

[87] The full reference is given in the next note. Thomas, on this issue, did not seem to be deeply puzzled as he carried forward from Augustine. Indeed, Aristotle—and his own obvious lack of experience—led him astray when he struggled with feminine humanity's meaning. From conversations with Lonergan, I know that he did struggle with that meaning very personally. What, then, are we to think of his "*Qualification*" in the concluding section of his brilliant article? The third sentence of "*A Qualification*" reads: "The precise implications of this doctrine are not too clear." Were they clear to him, but left as a challenge to the reader? He rambles round the mesh and mess of Augustine and Aquinas, pointing to the need to tune orgasm to the symphony of the journey to "our

2. The Insight

There is a nice shock value in expressing the key insight to be conveyed in the words of that same essay of Lonergan, so here is its verbalization by him.

> Now towards this high goal of charity it is no small beginning in the weak and imperfect heart of fallen man to be startled by a beauty that shifts the center of appetition out of self; and such a shift is effected on the level of spontaneity by *erôs* leaping in through delighted eyes and establishing itself as unrest in absence and an imperious demand for company.[88]

The shock here is that Lonergan is writing about Christian marriage, but I am asking for a fuller reach towards the meaning of sexuality, towards a larger divine delight of being with the children of men,[89] towards a massive creative cognition—not, then, a recognition—of the allure[90] of sexuality within "the order of the universe"[91] so that, in later millennia, good will will will well and it will be true that "good will wills the order of the universe and so it wills with that order's dynamic joy and zeal."[92]

3. The Handing On

I have so far not talked of functional collaboration, although I am following the general 4-piece formula suggested for these contributions. But the title of this section brings us quite obviously to the topic. To whom am I handing on, typing about, this insight, or set of insights? At first insight, the answer may seem easy. You are reading this and recognize a problem named by Lonergan in those final pages of the article "Finality, Love, Marriage," and, further, you react in your own way to my suggested push forward from Lonergan.

A key feature in our sharing here is my invitation now, herenow, to pause over my remote meaning for the word "suggested." While I follow the general formula, I do so in a skimpy appendix, skimpily. To some readers, like Daniel Helminiak, my skimpy pointings are a confirmation of the fruit of their own long struggle to escape

---

eternal embrace with God in the beatific vision" (end of page 51). His Conclusion ends with an appeal for scrutiny, corrections, and developments. See the text, and the further comment in the editors' note x on pages 263–64, regarding "the author's wish." The appeal and the wish were quashed by a central power.

[88] Bernard Lonergan, "Finality, Love, Marriage," *Collection, CWL* 4, 31–32.

[89] *Proverbs,* 8:31.

[90] See note 86 above.

[91] *Insight, CWL* 3, 722, end.

[92] *Ibid.*

Catholic madness.[93] But for many there is a huge psychic block. Harry Stack Sullivan writes in *The Interpersonal Theory of Psychiatry* of a parent's horror at their little girl's attention to her genitals: did they birth a monster? Was it such a parental horror in Rome regarding Lonergan's 1943 attention to our genitals that cried "hands off" in 1944? It will take more than a little scandalous appendix to shift the millennia of neurotic madness.

So I have not talked of this appended effort in terms of, say, a functional interpretation. It will take many full functional cyclings and re-cyclings[94] to lift us to the religious experiences of the *Kama Sutra*, to lead us to "view gender / As a beautiful animal / That people often take for a walk on a leash / And might enter in some odd contest / to try to win strange prizes."[95]

4. Further Context

My appendix is tagged on as a comment on footnote 83 of my Epilogue, where I mentioned a prayerful presence of sexuality. The prayerful presence in question is a kataphatic presence, not then some mystical achievement.[96] Yes, we are talking about the *Ascent of Mount Carmel*,[97] where Elijah's servant first found nothing, but, after seven climbs, saw "a little cloud no bigger than a person's hand, rising out of the sea."[98] Yet John of the Cross insists on saying "nada," seven times. For me the

---

[93] Daniel Helminiak's climb is laced, with exceptional openness and honesty, through his book, *The Transcended Christian: What Do You Do When You Outgrow Your Religion?* (Atlanta, GA: CreateSpace, 2013).

[94] The **pretend** of this volume of essays is that we share a communal Standard Model that hovers round the meaning of the symbol W_3. It will take many cycles in this century to reveal the depth of the pretense.

[95] Hafiz, "How Does It Feel to be a Heart?" quoted from Daniel Ladinsky, *I Heard God Laughing: Renderings of Hafiz* (New York: Penguin, 2006), 36.

[96] I need hardly here draw further attention to my life-long appeal to the cultivation of a contemplation that is essentially an incarnately thinking effort. But it is worth recalling my more recent emphasis on the luminous use of analogy in that thinking. Affirmation, negation, and eminence as a poise gives a climbing clarity to our lives with our Infinite Friends when affirmation dominates our conversations with Them, our searchings into Their Presence with us. The darkness of eminence is focused: see *CWL* 11, Thesis 5. Recall the challenge of "Foundational Prayer" that was the topic of the five essays *Humus* 4-8.

[97] The italics recall John of the Cross's work of 1578–79, which directs the searcher in a different way than the way talked of in the previous note. But there is, at times, in his writings, a sexual overtone that invites incarnate openness. So, the sixth stanza of the opening poem reads, can be read breastfully, "Upon my flowery breast, / Kept wholly for himself alone, / There he stayed sleeping, and I caressed him, / And the fanning of the cedars made a breeze."

[98] I *Kings* 18:44.

little cloud, after more than seven climbs, indeed more than seven times seven years of twisted climbing, finds newly the little cloud of *Insight* 691, "In the twenty -sixth place, God is personal." Am I freshly back, or rather forward, to the seeing and the seizing of the beginning of my Epilogue?

Have you climbed once, slowly?[99] Then I say, as Elijah said to Ahab, "Go again seven times,"[100] and you will meet not *nada* but **Nadia**. That 26th place can be a new open-nerve to a seeing of Thomas' twenty-seventh place.[101] But you must find your own new name, a mouth-stone twisting *Revelation*'s promise.[102] **Nadia**? It may only be personal to me, though its meaning in Russian and related languages is, providentially, **hope**, with **calling** as an Arabic meaning.[103] But for me the name leaps out from my Gaelic version of the 26th place, *Dia pearsanta*.[104] But how personal, *pearsanta*, is *Dia*? The Gaelic there, if lifted by Faith to a plural, *na* brings *Na Dia*, **Nadia**: what should be monotheistically *An Dia* is lifted to be the grammatical-conflicting *Na Dia*, begging for the 27th place. But do not mind my Joycean ways.[105] "Go again, seven times" till your Clasped mind finds, in your

---

[99] Are you and I, here, in the 1833 Overture of *Method in Theology*, page 250? The issue is leaving the God of Abraham and of the philosophers behind and climbing with Einstein, eyes on the secondary intelligibles (*Insight*, 649–51) and poised for an altogether new Christoffel, Christ-offered, Tensor. See Lindsay and Margenau, *Foundations of Physics* (New York: Dover, 1957), 364.

[100] I *Kings* 18:43.

[101] I am referring, of course, to the 27th question of the *Summa Theologica*, *Prima Pars*, lifted into this new little cloud of God-grasping.

[102] *Revelations* 2:17. "To anyone who conquers, I will give some of the hidden manna, and I will give a white stone, and on the white stone is written a new name."

[103] A rather startling set of accidents, like the double meaning of "Double You Three," and in the bottom of the diagram of *W₃* you find clues in my early struggling with a Trinitarian theology of history: the Father is associated with hope and named *Attractor*.

[104] *Dia* is the Gaelic word for God. It is not here, accompanied with the word 'an' (pronounced *un*), but it can be: thus, 'the god' is *an Dia*. The next sentences in my text point to the curious twist of giving the plural for 'the'—na (pronounced *gna*)—without shifting to the plural in Gaelic for 'gods.' *Na Dia* is just bad grammar. It neatly mixes plurality with singularity.

[105] Still, I would ask you to mind, mind-mind, the profound existential side of my language problem. Thomas puzzled over naming the Holy Spirit, but there was no problem, for him, with naming the First Person *Father*, as Jesus did. I have, here and there, raised problems regarding that naming, as feminists have, as neglected or abused children need to. The deeper issue is the precise meaning of *generation* when the generator, so to speak, emerges amoeba-wise, with the generated. Not, then, a Christian sublation of myths like those of Castor and Pollux, but the fresh directional mind-meaning of the Field-search struggled towards in *Verbum: Word and Idea in Aquinas, CWL* 2. And a step towards the coming convergence of global religiosity: the Beloved of high religious reaching is the Hope of the world.

neuro-molecules, a name that brings to you, in your little cloud of that First Person's loveliness, the lusciousness of the *Song of Song*'s Beloved.

Robert Henman, C.T.C., B.A., M.A. has been a lecturer in philosophy and family studies at Mount St. Vincent University, Halifax, Canada, from 1984 to the present. He has also lectured in medical ethics at Dalhousie Medical School, Dalhousie University. He has recently published *Global Collaboration: Neuroscience as Paradigmatic* (Vancouver: Axial Publishing, 2016) and, before that, *The Child as Quest* (Washington, DC: University Press of America, 1984). He has published articles on ethics, meta-neuroscience, psychotherapy, and theology. He is the General Secretary for SGEME: The Society for the Globalization of Effective Methods of Evolving. See http://www.sgeme.org/index.aspx. His current interest concerns genetic method in neuroscience. He may be reached at Robert.Henman@msvu.ca.

Terrance Quinn did his Master's and Doctoral work in C*-algebras, an area of mathematics originating in quantum physics. He held post-doctoral positions at Trinity College Dublin (1992–93) and University College Cork (1993–95). He has taught at Texas A&M International University (1995–2001), Ohio University Southern (2001–06), and Middle Tennessee State University, where he is currently professor and former chair of the Department of Mathematical Sciences. He has published in mathematics, mathematics pedagogy, and philosophy of science. Quinn's present work focuses on the writings of Bernard Lonergan as illuminated by Philip McShane. Quinn is gradually expanding his range to include religious studies. Book manuscripts presently in progress include *Invitation to Generalized Empirical Method in the Sciences* and *The (Pre-) Dawning of Functional Collaboration in Physics*. Currently Quinn's interests include studying and promoting functional collaboration.

Patrick Brown is an Affiliated Scholar with the Seattle University School of Law. He taught in the Seattle University philosophy department before joining the law faculty in 2002. He obtained his Ph.D. in philosophy at Boston College, writing his dissertation on Bernard Lonergan, and his J.D. degree at the University of Washington. Following law school, he clerked for the Chief Justice of the Washington Supreme Court and practiced law full-time for several years before beginning his teaching career. He has published articles in *Theological Studies*, the *Journal of Catholic Social Thought*, METHOD: *Journal of Lonergan Studies*, and the *Seattle University Law Review*, among other journals. With Michael Shute, he is currently coediting *The Selected Letters of Bernard Lonergan*. He may be reached at brownp@seattleu.edu.

Robert Aaron Mundine has worked for the last five-odd years studying the Greek achievement: the emergence of the theoretic lifestyle. His hope has been to raise awareness of the need for an integration of the human good on the level of historical

247

consciousness, where such awareness includes a serious grasp of what theory means and what a real apprehension of the concrete means. So far, so good: he has two completed works: "An Interpretation of Aristotle's *On the Soul* 429a 10–430a 25" and "Geometry as a Precondition to Interpreting the *Nichomachean Ethics*." He has an additional work in progress: "The Question of Priority in Methodological Thinking: Aristotle? Plato? Heraclitus?" Aaron can be contacted at aaron.mundine@outlook.com.

James Duffy obtained his Ph.D. in philosophy at Fordham University, writing his dissertation on Bernard Lonergan's two studies of Aquinas. Currently he resides in Morelia, Michoacán, Mexico, where he teaches English and philosophy at the *Instituto Tecnológico de Estudios Superiores de Monterrey* (ITESM). He has also taught and directed theses in pedagogy at the *Universidad Nova Spania* in Morelia. From 2006 to 2007 he collaborated in the design of a humanities and social sciences major at the ITESM. Prior to moving to Mexico, Professor Duffy taught philosophy, theology, and interdisciplinary studies to undergraduates and seminarians for five years at Saint Mary's University of Minnesota. Currently he is writing a book about appropriating lyrics of some favorite songs. He may be contacted at james.duffy@itesm.mx.

Bruce Anderson's research and writing is concerned with legal decision-making and Lonergan's economic theory. He is the co-author, with Philip McShane, of *Beyond Establishment Economics* (Halifax: Axial Press, 2002). Currently, his interest is method in law and legal theory. He is Professor of Business Law at the Sobey School of Business, Saint Mary's University, Halifax, Nova Scotia. He may be contacted at bruce.anderson@smu.ca.

Sean McNelis holds a bachelor's degree with honors from La Trobe University, a bachelor's degree in theology from Melbourne College of Divinity, and master's and doctoral degrees from Swinburne University of Technology in Melbourne, Australia. He has over 30 years' experience in housing management, housing policy, and housing research. He has worked as a housing manager living on a high-rise public housing estate, and he has also worked as a housing policy worker and advocate for low-income housing. At Swinburne he has undertaken many housing research projects for the Australian Housing and Urban Research Institute. In 2014 he published *Making Progress in Housing: A Framework for Collaborative Research* (Routledge) which draws on the writings of Bernard Lonergan. It presents a new approach to housing research, one that is relevant to all the social sciences.

William J. Zanardi has been on the faculty of St. Edward's University in Austin, Texas for several decades. He has specialized in the works of Bernard Lonergan and Eric Voegelin. His recent publications, some co-authored, include books on ethics (*Improving Moral Decision-Making*), liberty (*A Theory of Ordered Liberty*), and a four-volume series on applying Lonergan's fourth functional specialty, dialectic. He may be contacted at williamz@stedwards.edu.

Philip McShane is Professor Emeritus of Philosophy of Mt. St. Vincent University Halifax, Nova Scotia, Canada. He has written more than 20 books on philosophy, economics, and theology. A good deal of his writings are available on his website, www.philipmcshane.org, where there is an ongoing consideration of these issues. He has also edited two of the volumes of the *Collected Works of Bernard Lonergan*: *For a New Political Economy* and *Phenomenology and Logic*. McShane's most recent work, *The Allure of the Compelling Genius of History* (Vancouver: Axial Publishing, 2015), integrates Lonergan's two major works, *Insight* and *Method in Theology*, with the call of Jesus. He may be contacted at: pmcshane@shaw.ca.

Michael Shute teaches at Memorial University of Newfoundland. He writes on Lonergan's economics and is the author of, among other books, *Lonergan's Discovery of the Science of Economics* and *Lonergan's Early Economic Research* (University of Toronto Press, 2010). He is currently working on an edition of Lonergan's letters with Patrick Brown.

Michael George studies philosophy, history, and religion and obtained a Ph.D. from St. Paul University and the University of Ottawa on the topic of ethics and imagination. Since 1988, he has taught in the Religious Studies Department at St. Thomas University, in Fredericton, New Brunswick. He is currently Chair of the department and the Coordinator of Interdisciplinary Studies. His interest in ethics includes theoretical and applied issues. Ongoing projects include media and consciousness, bioethics, human sexuality, and obstacles that hinder the development of coherent ethical foundations. Sources of hope and insight include children, grandchildren, music, novels, and poetry.

Meghan Allerton is a research associate at the University of Toronto Scarborough Ecological Modelling Lab, where she is involved in developing a watershed modelling strategy to improve nutrient loading estimates for the southeastern Georgian Bay area. She is a graduate of the Master of Environmental Science program at the University of Toronto Scarborough (2015), and holds an Honours B.Sc. in Ecology and Evolutionary Biology from the University of Toronto (2013). Her article, "Empirical Exercise: The Dynamics of Knowing," appeared in the 2015 issue of the *Journal of Macrodynamic Analysis*. She can be reached at meghan. allerton@alum.utoronto.ca.

Made in the USA
Middletown, DE
15 December 2022

18024177R00156